Map generalization

Map generalization:
Making rules for knowledge representation

Edited by

Barbara P. Buttenfield
National Center for Geographic Information and Analysis
Department of Geography, SUNY-Buffalo

Robert B. McMaster
Department of Geography, University of Minnesota

Foreword by

Herbert Freeman
Professor of Computer Engineering, and
Director, Machine Vision Laboratory,
CAIP Center, Rutgers University,
Piscataway, New Jersey

Longman
Scientific &
Technical

Longman Scientific & Technical,
Longman Group UK
Longman House, Burnt Mill, Harlow,
Essex CM20 2JE, England
and Associated Companies throughout the world.

Copublished in the United States with
John Wiley & Sons, Inc., 605 Third Avenue, New York, NY 10158

First published 1991

Trademarks
Throughout this book trademarked names are used. Rather than put a trademark symbol in every occurrence of a trademarked name, we state that we are using the names only in an editorial fashion and to the benefit of the trademark owner with no intention of infringement of the trademark.

British Library Cataloguing in Publication Data
Map generalization: Making rules for
knowledge representation.
  I. Buttenfield, Barbara P.
  II. McMaster, Robert B.
  526.028

  ISBN 0–582–08062–2

Library of Congress Cataloging-in-Publication Data
Map generalization : making rules for knowledge representation /
  editors, Barbara P. Buttenfield, Robert B. McMaster ; foreword by
  Herbert Freeman.
    p. cm.
  Articles based on papers delivered at a symposium held during
April 1990 in Syracuse, N.Y.
  Includes bibliographical references and index.
  ISBN 0–470–21803–7
  1. Cartography—Congresses. I. Buttenfield, Barbara Pfeil.
II. McMaster, Robert Brainerd.
GA101.2.M36 1991
526—dc20   91–22294
                                                        CIP

Set in Times 10/12 point by Columns of Reading.

Printed and bound in Great Britain at the Bath Press, Avon.

To George F. Jenks and John C. Sherman
for their insight and guidance on
issues of generalization and design

Contents

Contents

List of contributors

Prof. Marc P. Armstrong
Department of Geography, University of Iowa, USA

Prof. M. Kate Beard
NCGIA, Department Surveying Engineering, University of
Maine-Orono, USA

Prof. Barbara P. Buttenfield
NCGIA, Department of Geography, SUNY-Buffalo, USA

Dr Gail Langran
Intergraph Corporation, Reston, Virginia, USA

Dr William Mackaness
Private consultant, London, UK

Prof. David M. Mark
NCGIA, Department of Geography, SUNY-Buffalo, USA

Prof. Robert B. McMaster
Department of Geography, University of Minnesota-Minneapolis, USA

Prof. Mark Monmonier
Department of Geography, Syracuse University, USA

Prof. Dr Jean-Claude Muller
Department of Cartography, ITC, The Netherlands

Prof. Bradford Nickerson
School of Computer Science, University of New Brunswick, Canada

Prof. Timothy Nyerges
Department of Geography, University of Washington-Seattle, USA

List of contributors

Ms Diane Richardson
Canada Centre for Mapping, Energy, Mines and Resources, Ottawa, Canada

Mr K. Stuart Shea
TASC, Reston, Virginia, USA

Dr Robert Weibel
Department of Geography, University of Zurich, Switzerland

Foreword

A map is an image that serves as a medium of communication, conveying spatial relationships to its viewer. It is an abstraction, or generalization of physical reality, and its effectiveness as a communication medium is strongly influenced by the nature of the spatial data, the form and structure of the representation, the intended purpose, the experience of the viewer, and the context in time and space in which the map is viewed.

This book deals with the problem of automating the process of generalization in map production. Map generalization is a tedious task, requiring skilled cartographers working for long periods of time. It is the common wisdom today that such labour-intensive tasks should be consigned to computers and thus be accomplished more uniformly, more precisely, more rapidly, and at much reduced cost. However, this has been difficult to achieve. Generalization is complex, and much of its execution is left to the *ad hoc* decisions of the experienced cartographer. The problem has much in common with other ill-defined tasks that humans have performed well for decades but that now beg to be solved more rapidly and consistently in an automatic fashion.

The expert system approach has been put forward as a promising route for accomplishing this: systematically capturing the collective lore used by humans in performing the task, classifying and organizing this into an explicit list of rules (the rule base), and then applying it by means of a logical-inference program to the task at hand. The approach has helped to provide structure for solving these problems, and has yielded good results in a number of cases: engine-fault analysis, oil exploration, and in cartography, automatic name placement.

Precise, comprehensive understanding of a problem, followed by explicit documentation, has been the foundation of all of mankind's technological accomplishments, and there is no question but that the development of a rule base is central to the ultimate solution of this problem, though there may be some differences of opinion as to the specific route to follow.

Buttenfield and McMaster have assembled here a collection of 13 chapters dealing with the development of a rule base for map generalization. The chapters are based on papers delivered at a three-day symposium held during April 1990 in Syracuse, New York, and attended by most of the leading researchers in automatic map generalization.

The book represents an in-depth examination of the current status of automated map generalization – the accomplishments to date, the current issues, and the long-term challenges. It is impressive both in the depth to which the issues are examined and in the breadth of its coverage. It should be recommended to all who have an interest in automated cartography, as well as to those computer scientists and computer engineers who are ever looking for applications in which to bring the information processing power of the digital computer to bear on human tasks that are tedious and demanding of high professional skills. For cartography, the benefits will be improved accuracy, greater consistency, and reduced production time.

Herbert Freeman
Rutgers University

Preface

This research volume focuses on formalization of cartographic knowledge for digital map generalization. Up to this point in time in the development of cartographic research, it has been commonly accepted that much of the cartographic process is accomplished intuitively. Map compilation and design standards, as well as most generalization operations, have been established in large part by trial and error, and by repetition. In some cases, empirical evaluation of specific design criteria (e.g. graduated circle scaling, development of grey-tone progressions) has been carried out through psychophysical testing techniques, but these aspects have been tested in isolation. It is difficult to design psychophysical research using embedded information. Unfortunately, most of the empirical research has not been carried out in electronic mapping situations, nor in the context of GIS analysis and decision-making. Nowhere is this issue more evident than in the area of map generalization. Training in cartographic skills often proceeds by the rule of 'this looks about right', despite sparse attempts by cartographers (such as Richard Edes Harrison and George Jenks) to offer training in 'logical generalization' utilizing manual techniques.

Formalism in the area of map generalization remains an unsolved problem, and this has created obvious impediments to automating the cartographic process. Cartographic research in the past decade has emphasized algorithm development and assessment, error determination, formal description of map feature geometry, and the development of logical models. Three epochs of research in automated generalization may be identified. During the first, from approximately 1960 to 1975, research focused upon algorithm development, with particular emphasis on algorithms for line simplification. During the period, emphasis on processing and smoothing raster images was also prevalent (take for example the development of SYMAP and several of its offshoots). Emphasis later in the period on vector representations and topological data structures was synchronous with refinements in electronic archival and display technology. In the second epoch, during the late 1970s and the 1980s, assessment of algorithm efficiency became an increasing concern. The scale-dependent nature of geographical phenomena was modelled using parametric and self-similar methods. Unfortunately, most research viewed phenomena in isolation, ignoring the need to integrate generalization procedures.

A third epoch in map generalization research has recently become apparent, and continues to develop in many disciplines. Formalization of cartographic knowledge is being explored through development of 'comprehensive' models, application of expert systems and knowledge-based techniques. Full automation of the generalization process still lies some years ahead. The generation of digital cartographic data from multiple data sources and formats for representation and analysis at multiple scales continues to be an issue of theoretical and pragmatic interest. Implementation of a formalized rule base to guide the mapping process may improve efficiency, preserve consistency, and incorporate sound principles in digital mapping. A recent research initiative ('Multiple Representations') funded by the National Center for Geographic Information and Analysis (NCGIA) created a forum for international discussion and resolution of impediments associated with formalizing cartographic knowledge and automating generalization.

The editors of this volume decided to pursue the generalization problem in a subsequent research symposium 'Towards a Rule Base for Map Generalization', funded jointly by Syracuse University and the NCGIA, and held in Syracuse in April 1990. Fifteen researchers from Europe and North America attended to discuss impediments in acquisition and representation of cartographic knowledge. Impediments may be found in data abstraction, data archival and management, and in data representation. Rudimentary rules have been codified. Examples range from the very simple Radical Law published by Töpfer and Pillewizer (1966), to comprehensive attempts such as the compilation of rule bases developed by the US Defense Mapping Agency. Other examples can be readily discovered, but there is little consistency in rule formation or application, making integration and co-ordination difficult. Additionally, the intuitive aspects of map generalization have not been addressed or codified.

Following the symposium, research to address these impediments was pursued; participants corresponded and groups of individuals met and reported on their research progress at conferences throughout the remainder of 1990. The conferences included the Association of American Geographers' meeting in Toronto, the 4th International Symposium of Spatial Data Handling in Zurich, and the GIS/LIS meetings in Anaheim, California. The editors would like to acknowledge the travel funding provided to participants by their respective firms and universities. This book integrates the ideas and threads running through the body of research that has developed. This type of long-term co-ordination of research efforts is relatively rare, and the integration of ideas coming from participant efforts over the past months demonstrates the benefits of a truly integrated multidisciplinary effort. The intention in publishing this research volume is to overcome impediments to the formalization of cartographic knowledge, and to promote a continuing formum in the research community.

The book includes four parts: rule base organization; data modelling issues; formulation of rules; and computational and representational issues. The editors have invited perspectives from disciplines of cartography, computer science, and surveying engineering, to present a broad perspective on the formalization of guidelines for digital map generalization. These chapters go beyond a review of research results both to prioritize a research agenda and identify which aspects of the map generalization problem may be addressed immediately, given present knowledge and technology.

Throughout this book, terminology proposed by the National Committee for Digital Cartographic Data Standards (NCDCDS 1988) will be used. The National Committee advocated standard usage of definitions and terminology, and the editors have adopted those standards wherever possible. For example, readers will encounter the tripartite terminology of 'entity' in the world, 'object' in the database, and 'feature' as the union of an entity and one or more objects. Where important distinctions must be made (for example in the use of the term 'object orientation', which is not included in the proposed standard) the editors have diverged from this document. In similar fashion, the adoption of the term 'digital generalization' is intended to present a broad definition incorporating numerical, symbolic, and graphical manipulations of map data. Adoption of these broad standards is intended to integrate the diverse perspectives presented in this volume, and to co-ordinate them in a common framework.

Acknowledgements

The editors are most grateful to institutions and individuals for their assistance. Special thanks go to Robert G. Jensen (currently Dean of the Graduate School at Syracuse) and Ross D. MacKinnon (currently Dean of the Faculty of Social Sciences at SUNY-Buffalo) for their support with all facets of the project and their steady confidence in two junior faculty. We received valuable advice at critical junctures from David Mark (SUNY-Buffalo), and Mark Monmonier and Dan Griffith (Syracuse). We are grateful to Herb Freeman for contributing his perspectives during the symposium and for his contribution of a foreword to the book. Ren Vasiliev (Syracuse University) served as rapporteur and managed local arrangements for the symposium, which simply could not have run smoothly without her hard work.

All chapters in this book have undergone a complete one-way blind peer review, just as for refereed journal publication. The book is better for the review process, and the authors are to be commended for their timely and agreeable acceptance of revisions. Reviewers included the following individuals (in alphabetical order): Bill Carstensen, Nick Chrisman, Keith Clarke, Bob Cromley, Joe DeLotto, Paul Densham, David Douglas, Geoff Dutton, Ron Eastman, Pete Fisher, Ann Goulette, Christopher Jones, Elizabeth Kohlenberg, Werner Kuhn, Alan MacEachren, Matt McGranaghan, Barbara Petchenik, Tom Poiker, Bernd Powitz, Charlie Schwarz, Terry Slocum, Khagendra Thapa, Marv White, and Pinhas Yoeli.

Barbara P. Buttenfield and Robert B. McMaster
February 1991

Part I

Rule base organization

1

Design considerations for an artificially intelligent system

K. Stuart Shea

Introduction

One of the most intellectually and technically challenging aspects of the mapping process is cartographic generalization. As the mapping process transits to a digital environment, it is vital to embrace a generalization philosophy which meets the changing needs of that transition, and resists merely automating established manual generalization procedures. The entirety of the digital generalization process must be evaluated by looking at the interrelationships between the conditions that indicate a need for its application, the objectives or goals of the process, and the specific spatial and attribute transformations required to effect the changes (Shea and McMaster 1989).

The manual generalization process is subjective, interactive, idiosyncratic and comprehensive in its perception and execution. By design, an enigmatic process is being automated. The complexity of generalization decisions demands that cartographers intelligently exploit their innate understanding of geographical phenomena to create a representation consistent with existing geographical knowledge. The lack of this innate, intelligent understanding has so far hampered efforts to automate the generalization process. More precisely, it is a limited understanding of how to use and apply geographical knowledge that has limited success. To intelligently integrate that knowledge in a digital environment, algorithm designs, implementation strategies, and control techniques for generalization operations must consider the geographical implications of generalization decisions.

This chapter will explore several design considerations for an artificially intelligent digital generalization system. Limitations of previous automation attempts are addressed first, followed by an examination of expert systems development. The structure of expert systems is reviewed, and several design considerations for a digital generalization system are covered. A proposed strategy to formalize the knowledge within a production rule-based system concludes the discussion.

Automating the generalization process

Historical perspective

Over the last 25 years, academic and industry researchers alike have noted the difficulties in automating the generalization process. Many of those researchers are still not today convinced it is possible. Eduard Imhof (1982: 357–8) perhaps articulated it best when he asserted: 'the content and graphical structure of a complex, demanding map image can never be rendered in a completely automatic way. Machines, equipment, electronic brains posses [*sic*] neither geographical judgment nor graphic–aesthetic sensitivity. Thus the content and graphic creation remain essentially reserved for the critical work of the compiler and drawer of a map.'

Some authors, such as Keates (1973), though less convinced of the professed futility of automation, are still reticent about presenting the possibilities, and simply offer cursory treatment of the topic. The most common sentiment purports that automated generalization is merely elusive (Robinson, Sale, and Morrison 1978). Brophy (1972: 8), suggests that this is due to 'a consequence of the ambiguous, creative nature . . . which lacks definitive rules, guidelines, or systemization'. None the less, many have endeavoured to automate various aspects of the generalization process.

Theoretical work on map generalization by Perkal (1966) and Tobler (1966) established the foundation for future efforts in digital generalization. Many others extended these initial efforts by primarily focusing on the generalization of linear digital data (Deveau 1985; Dettori and Falcidieno 1982; Jenks 1981; Douglas and Peucker 1973; Boyle 1970; Lang 1969). Though many of these efforts helped to establish several algorithms for conducting linear generalization, other authors have focused on the identification of the appropriateness of algorithm selection (McMaster 1983, 1986, 1987), and the relationship of the algorithm's point selection techniques to that of perceptual criticality (Jenks 1985; White 1985; Marino 1979). Greatly overshadowed by the work on linear data, the generalization of point and area features has also been addressed by several authors, but has resulted in fewer significant achievements (Monmonier 1983; Chrisman 1983; Lichtner 1978; Töpfer and Pillewizer 1966).

Generalization in isolation and abstraction

Many authors have repeatedly emphasized that manual generalization not be conducted in isolation or in the abstract (Robinson, Sale, and Morrison 1978; Raisz 1962), yet the early attempts at automation cited above often disregarded that guidance. For example, several early attempts at automation targeted a particular generalization process (such as the selection of point features), and addressed it partially, if not completely, in isolation from other generalization decisions (Catlow and Du 1984; Chrisman 1983;

Lichtner 1979). Moreover, development of these generalization techniques rarely considered the underlying geographical significance of the features, and often performed generalization operations on abstract graphic entities. Linear simplification activities are an ideal example (Vanicek and Woolnough 1975; Gottschalk 1973; Douglas and Peucker 1973; Boyle 1970; Maling 1968). In the case of linear simplification, only of late have authors addressed the issues of isolation and abstraction. McMaster (1989b) recently investigated the integration of heretofore isolated simplification and smoothing algorithms, and has shown some promising results. The preservation of the geographical process (Mark 1989) and interrelationships between features during scale change (Monmonier 1989a) have explored the issues of abstract generalization, and provide strong arguments for considering the geographical implications of generalization decisions.

Digital processing capabilities have been partially to blame for the focused examination of isolated and abstract operations in generalization. Digitization processes have provided spatial objects – digital representations of geographical phenomena – which often lack indication of source, scale, feature characteristics, or other specific attribution. Use of serial logic computers and conventional high-order programming languages has necessitated the independent treatment and sequential, predetermined application of generalization manipulations on those spatial objects. This approach is inconsistent with the manual generalization process wherein many generalization decisions are simultaneous, rather than sequential (Caldwell, Zoraster, and Hugus 1984). In the manual process, knowledge of the geographical relevance of generalization decisions supports this simultaneous processing methodology. The ability to exploit existing computing technology to perceive the map as a whole does not yet exist. As a result, we have not yet acquired the ability to instruct the computer to assess the impact of generalization decisions made for one feature upon another feature.

Generalization in a digital environment

Many of the traditional manual generalization processes will be necessary in digital applications, therefore the selection and order of these processes become as crucial as the extent or limit of their application. Recently, several authors have proposed models for the digital generalization processes (McMaster and Shea 1988; Brassel and Weibel 1988). These models provide a basis which furthers an understanding of the complicated and interrelated aspects of generalization processes. Ultimately, these processes will represent a series of operational steps in a digital environment. Several key issues, some seemingly unresolvable, challenge attempts to emulate digitally this traditionally manual process.

Current attempts at automating generalization operations lack the means to incorporate the traditional cartographic intuition – the intellectual basis provided by the cartographer – when confronted with a generalization problem. Recent advances in the field of artificial intelligence (AI) offer exciting possibilities to assist in this endeavour. The concept of an artificially intelligent digital generalization system suggests methodologies which reflect the theoretical aspects of human intelligence and, as such, could closely mimic the human generalization process. A judicious application of the concepts and techniques promulgated in the AI field may serve to bring the intuitive basis of generalization to the digital environment.

Artifical intelligence

Historical perspective

Although the conceptual basis of AI traces back to Aristotle (Sowa 1984), much of what is known today has its roots in the period from 1930 to 1950 when AI research efforts were focusing on formalized reasoning using logical systems and the development of large computational devices (Smith 1984). In the 1950s AI research began to explore the fundamentals of symbolic (or non-numerical) reasoning. It was quickly discovered that an ill-defined intellectual component was necessary for this form of reasoning. The 1960s brought about efforts to simulate the complicated process of thinking by finding general methods for solving broad classes of problems. These efforts proved unsuccessful, however, since programs which con-tained general problem-solving procedures and possessing little knowledge specific to the domain under investigation were often inefficient at problem-solving. Success was finally nearing in the 1970s as researchers began to consider methods of capturing and representing knowledge in their computer programs.

In 1977, Edward A. Feigenbaum altered the focus of research in the AI field dramatically by noting that the problem-solving power of a computer program is derived from the knowledge it possesses, not from the particular formalism and inference schemes it employs. This realization launched the development of a branch of AI known as **expert systems**. Expert systems are computer programs that manipulate symbolic knowledge and heuristics to simulate human experts in solving real-world problems (Weiss and Kulikowski 1984). More specifically, expert systems are 'capable of representing and reasoning about some knowledge-rich domain . . . with a view to solving problems and giving advice' (Jackson 1986: 1). Thus, expert systems not only embody expert knowledge, they also have the ability to recount the steps taken to solve a problem, as well as gain proficiency at a particular task.

Applications of expert systems

Expert systems have recently received a considerable amount of attention in the cartographic literature, with applications to automated name placement and conflict detection (Pfefferkorn *et al.* 1985; Freeman and Ahn 1984), computer-assisted map design (Muller, Johnson, and Vanzella 1986; Robinson and Jackson 1985), automatic generalization (Nickerson and Freeman 1986), as well as general treatises on several other possible applications (Ripple and Ulshoefer 1987; Mackaness, Fisher, and Wilkinson 1986; Graklanoff 1985). The near-term developments in expert systems for cartography indicate a strong developmental future, although some question the cartographic validity of these systems (Fisher and Mackaness 1987).

A major stumbling block in these expert system developments is the formalization of the cartographic knowledge needed to address a specific problem (Robinson *et al.* 1986). This barrier is not unlike that which already confronts the automation of any intellectual experience. Making generalization decisions requires intelligence, and a lack of this intelligent character has hampered previous efforts to automate the generalization process. Moreover, there is a limited understanding of how to formalize and apply the knowledge at our disposal. To fully appreciate the problems of formalizing that knowledge, the nature of expert systems, and the role that knowledge plays within them, must first be understood. The following review on the organization of expert systems seeks to establish such a basis of understanding. Then we can begin to estimate the possibilities of applying this technology to generalization decisions, with an ultimate goal of automating significant aspects of the digital generalization process.

Organization of expert systems

The components of a generic expert system are shown in Fig. 1.1, and include: (1) **a knowledge base**, which contains a representation of the knowledge that is required to assess relations between entities in a database; (2) **an inference engine**, the means by which this knowledge is handled; (3) **the object database**, a set of facts reported or inferred about the problem at hand by the expert system; and (4) **the input/output user interface**, which enables the user to supply facts and data, and enables the system to ask questions or supply advice and explanation.

The **knowledge base** is derived from operating procedures, facts, concepts, and principles standard in a specific domain. The knowledge base may contain knowledge of two basic types: declarative and procedural. Declarative knowledge includes facts, concepts, and descriptive information about a particular domain (not unlike the textbook knowledge of a field). Procedural knowledge covers the procedures, strategies, and heuristics employed to solve problems or to achieve intended goals (the 'how to' aspect of the domain). Feigenbaum (1977) calls these two types of

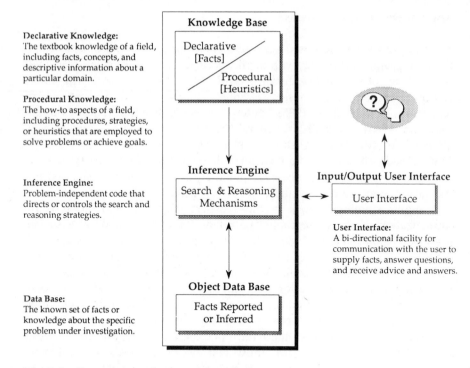

Declarative Knowledge:
The textbook knowledge of a field, including facts, concepts, and descriptive information about a particular domain.

Procedural Knowledge:
The how-to aspects of a field, including procedures, strategies, or heuristics that are employed to solve problems or achieve goals.

Inference Engine:
Problem-independent code that directs or controls the search and reasoning strategies.

Data Base:
The known set of facts or knowledge about the specific problem under investigation.

Knowledge Base

Declarative [Facts]

Procedural [Heuristics]

Inference Engine

Search & Reasoning Mechanisms

Object Data Base

Facts Reported or Inferred

Input/Output User Interface

User Interface

User Interface:
A bi-directional facility for communication with the user to supply facts, answer questions, and receive advice and answers.

Fig. 1.1 Components of an expert system

knowledge 'facts' and 'heuristics'.

The handling of knowledge requires intelligence. A measure of intelligence includes not only what is known, but also how well that knowledge can be used and applied (Hart 1986). The strength of an expert system lies in the combination of these two important features. Therefore, access to a wealth of knowledge is useless without an understanding of how to select, and when to apply, the appropriate knowledge to the problem. The **inference engine** provides this control mechanism to direct the search and reasoning processes. In effect, an inference engine drives an expert system. It controls the search, selection, and manipulation of data in a knowledge base in finding an acceptable solution. A key element of an expert system is the separation of knowledge about the problem domain, from the knowledge which understands how to solve problems. In fact, a key criterion for an inference engine is that the inference or control mechanisms function independent of the facts of the domain, while still remaining adaptable to a divergent set of domain facts.

The **object database** contains information related to the specific processing problem at hand, and may be built from facts supplied interactively by the user, or can contain information derived by the program as part of its processing. Consisting of two parts, the current task-domain situation and the goal, the object database provides the assertions or facts which represent the current problem, and the goal condition against which those facts are directed.

The **input/output user interface** enables both the system designer and user to communicate with the expert system. It is this bidirectional communication facility that enables the user of the system to supply facts, answer questions, and receive advice and answers. Additionally, it enables the developer of the system to modify the knowledge base, its representation, or its storage methodology.

Design considerations for intelligent automation

Several important decisions must be made when designing an artificially intelligent digital generalization environment. These decisions generally focus on the principal elements of an expert system design noted above: the knowledge base, the inference mechanism, the object database, and the input/output user interface. Clearly, each decision depends on the problem domain under evaluation, as well as the specific operational intent of the system being developed. Of particular importance are two items: the development of a viable scheme for knowledge formalization, and the selection of appropriate search and reasoning strategies. Their critical importance in developing expert systems warrants a more detailed examination.

Knowledge representation

Given a lack of a strong theoretical understanding of the function and structure of maps and of the individual generalization processes, flexible data models and adaptive, intelligent algorithms necessary to formalize and exploit the problem-solving knowledge essential to generalization are currently lacking (Brassel and Weibel 1988). Concomitantly, of all the design decisions needed to make an expert system operate, the principal determinant of success is the sophistication and breadth of the contained knowledge. The generation and maintenance of the knowledge base containing the geographical and cartographic knowledge provide a basis for contextual reasoning. The knowledge engineering process, which builds the knowledge base, attempts to acquire from experts the procedures, strategies, and rules of thumb for problem-solving. Before that knowledge can be exploited, it must be formally structured according to the needs of the specific expert system.

The selection of a formalism for knowledge representation is a significant distinguishing characteristic when designing an expert system. There are several knowledge representation techniques, any of which can be used alone or in conjunction with others to build expert systems. Each of these knowledge representation formalisms has demonstrated a particular affinity for providing the program with certain benefits, such as making it more efficient, more easily understood, or more easily modified (Waterman 1986). No formal metric exists, however, to test the appropriateness of a particular representation scheme (Barr and Feigenbaum 1981). Nyerges (1991c, Ch. 4 this volume) reports on the suitability of these knowledge representation schemes for geographical information abstraction. Presented below are several of these representation techniques.

Production rules

A popular procedural system for knowledge representation is the production system. Expert systems that use production systems to represent knowledge are commonly referred to as rule-based systems. In rule-based systems, knowledge representation centres on the use of the rule syntax:

If <antecedent> : THEN <consequent>

This structure is termed a production rule, and is composed of the left-hand side antecedent (or predicate), a logical combination of propositions about the database and a separate right-hand side consequent, containing a collection of actions. The antecedent of a rule states a condition or aspect of the problem that must be present for the production to be applicable, while the consequent specifies an appropriate action to take. During the execution of a production system, a production rule is triggered when the antecedent of a rule is matched. A triggered rule is not executed, but is merely examined with all other triggered rules. When all rules have been examined, and the most appropriate rule is determined, that rule is then fired, resulting in the performance of the consequent of the rule.

In many rule-based systems, procedural knowledge in the form of heuristic IF–THEN production rules is integrated with declarative knowledge. These production rules, called **metarules**, help guide the execution of a rule-based system by determining under what conditions certain rules should be considered instead of other rules, or how certain rules should be modified. Metarules embody the control knowledge necessary for indicating the sequence to apply the available declarative or procedural knowledge in solving a problem. The following metarule is provided to illustrate:

IF   product scale is $\geq$ 50 000
THEN examine all rules relating to point features

The use of production rules offers several advantages in knowledge representation. First, knowledge in the form of production rules is extremely readable and easy to understand. Other knowledge representation schemes, such as semantic networks, use terminology that is often imprecise and cannot be parsed to give unambiguously correct relationships between objects. Second, an obvious quality of production rules is that they behave much like independent pieces of knowledge and, as such, rules in the knowledge base can be independently added, deleted, or modified with little direct effect on other rules.

Modification of the knowledge base is therefore relatively simple. Finally, an ordered list of the rules fired in the decision-making process helps to explain the line of reasoning or to justify conclusions reached. This last capability provides real advantages for generalization. Even the human cartographer has a limited ability to explain generalization decisions; this is well demonstrated by the historical difficulties in developing objective rules for generalization. As rules trigger, lists of all triggered rules – known as inference chains – can be easily compiled. The inference chains provide an audit trail to the user on how the system reached its conclusions, though not necessarily why.

Production rules do suffer from several disadvantages. The most significant disadvantage is that rules in a production system lack topology. The strong modularity of the productions results in inefficiencies of program execution, and makes rule maintenance unwieldy as the number of rules increases. Several other knowledge representation schemes address these deficiencies. Production rules may be expressed in several ways: as logic, structured in a semantic network, or they may be gathered together in a frame. These approaches for knowledge representation are useful for representing facts about objects and how those objects relate to each other.

Logic

Many expert systems have been built around the use of predicate logic formalisms for representing knowledge. In this approach, a set of predicates, and the logical relationships between predicates, are specified. These logical relationships correspond primarily to the typical Boolean and conditional expressions (such as: AND, OR, IF–THEN). The predicate logic formalism permits expression of facts and rules. First-order predicate calculus and other similar formal logic languages have been used to represent declarative knowledge, such as the case wherein instantiated predicates (relations) represent facts (Gallaire, Minker, and Nicolas 1984; Mylopoulos 1980). Rules can be represented as 'predicate A if predicate B'. The programming language PROLOG (PROgramming in LOGic) uses such an approach (Kowalski 1979).

The use of logic as a technique for knowledge representation has been extended to procedural knowledge as well. In addition, the use of less formal logic to represent knowledge (e.g. fuzzy logic) is designed to handle concepts that are relative and approximate. These concepts are an anathema to formal logic languages, but are closely analogous to human thinking.

Advantages of logic representation are that logic: (1) represents a natural way to express an intuitive understanding of a domain; (2) is precise and consistent in the expression of the formalism; (3) offers flexibility in that the representation of fact is not tied to the use of that fact; and (4) includes logical assertions which are modular and independent from one another (Barr and Feigenbaum 1981). The most significant disadvantage is that no hierarchical framework of embedded rules can exist.

Semantic networks

Semantic networks, or nets, originally introduced as means of modelling human associative memory (Quillian 1968), have been extended to represent declarative knowledge. In structural terms, a semantic net consists of objects, represented as a set of nodes in a graph, with the relations among them represented by labelled arcs indicating the type of relationship (such as: a-kind-of, a-part-of) (Ballard and Brown 1982). These network structures permit simple deductions through property inheritance, items lower in the network inherit properties from items higher in the network. As a simple example, consider the statements 'Route 66 is a kind of interstate highway' and 'Every interstate highway is a kind of transportation feature'. To illustrate, these can be represented in a semantic network as shown in Fig. 1.2. This example uses an important type of arc: a-kind-of. Because the properties of the relations linking the nodes is known, it is possible to infer a third statement from the network, that 'Route 66 is a transportation feature', even though it was not explicitly stated.

Semantic networks are sometimes embedded in systems that also have other knowledge representation schemes. For domains that include a significant amount of knowledge with complex interrelationships, a semantic network can provide the foundation for a sophisticated inference system. A key feature of these networks is that explicit relations between objects allow implied relations to be derived without searching through the entire set of relations.

Frames

Frames comprise another common representation method for declarative knowledge (Minsky 1975). Frames are analogous to data structures where all declarative and procedural knowledge about a particular object or event

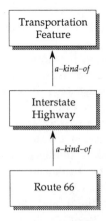

Fig. 1.2 A semantic network using the relation termed 'a-kind-of'

is stored together. A frame represents an object as a group of attributes, wherein the stereotyped 'image' of a particular object invokes a stereotyped 'image' of a set of attributes. This formalism facilitates expectation-driven processing. This allows all known knowledge to be readily accessible when a specific decision-making need arises.

Frames are organized much like semantic networks. A frame is a network of nodes and relations organized in a hierarchy, where the topmost nodes represent general concepts, and the lower nodes more specific instances of those concepts. Unlike a semantic network, in a frame system the concept at each node is defined by a collection of attributes and values of those attributes, where the attributes are called slots. Each frame has a name, and a list of slots, which have either values assigned to them, or procedures which are executed to produce a value (known as procedural attachments).

There are several advantages to using a frame-based formalism for knowledge representation. First, since both declarative and procedural knowledge are maintained, they support both top-down and bottom-up control strategies. Second, the structure can be expanded to incorporate other knowledge by linking additional frames, and the structure easily supports the incorporation of production rules (Armstrong 1991, Ch. 5 this volume).

Inference engine and search strategy design

Whatever formalism scheme is chosen, simply having access to a great deal of knowledge is not enough; there must be a mechanism that directs the implementation of the knowledge against the problem under consideration. The inference engine contains this logic to control and direct the search and reasoning techniques. In a production system, for example, the logic to control and direct the search techniques can be characterized as having four basic parts (Michaelsen, Michie, and Boulanger 1985), including: (1) selection of the relevant rules and data elements; (2) matching the active rules against data elements to determine which rules have been triggered, indicating they have satisfied the antecedent condition; (3) scheduling which triggered rules should be fired; and (4) execution (firing) of the rule chosen during the scheduling process. The choice of an appropriate control strategy to address these four actions is dictated by the problem under consideration, the content of the object database, and the strucure of the knowledge base. The control strategy uses the rules in the knowledge base to manipulate the data in the object database which defines the problem.

A simple inference engine would be theoretically sufficient for processing rule bases of any size; however, as the number of rules increases, acquiring all available evidence or primitive values would become very inefficient. In order to manage efficiently the application of knowledge to specific problems, inference engines apply a control strategy that carefully controls the order of rule activation, as well as controlling the search techniques used to determine how the rules are applied to the problem. Given an initial state, which defines the current task-domain situation, the inference engine attempts to control the processing to reach a goal state. This process of sifting through alternative solutions to proceed from the initial state to the goal state is called search.

Inference engines typically employ one of two basic types of inference methods for exploiting rules – forward chaining and backward chaining – for the route followed to form an inference of each type Hayes-Roth, Waterman, and Lenat 1983). In forward chaining the inference engine works forward from known or asserted predicates to derive as many consequents as possible (antecedent-driven). In essence, rules are matched against facts to establish new facts. In backward chaining, the inference engine works backwards from a hypothesized consequent to locate antecedents that support it (consequent-driven). In this inference method the system starts with what it wants to prove, and then tries to establish the facts it needs to prove the finding. Forward-chaining reasoning is a more appropriate choice for generalization since the process is data-driven and situation specific.

Search strategies, such as breadth-first, best-first, depth-first, uniform-cost, bidirectional, and heuristic searches, decide the paths and directions taken to find the rule or sequence of rules which match the conditions determined to help solve the problem. More importantly, these search strategies help to narrow the problem by determining efficiencies of goal satisfaction. Both sequential and iterative applications of the rules may be necessary, in addition to backtracking controls to revisit promising paths.

Finally, in many problem domains, knowledge is not known with certainty. Consequently, while the rules in a knowledge base may reflect logical conclusions, if the given prerequisite items of evidence are uncertain, so too must the resulting conclusion be uncertain. As a result, various formalisms have been developed that calculate a 'degree of belief' value for various possible conclusions. These techniques include: modified Bayesian techniques, based on Bayes' rule; evidential reasoning techniques, which provide a set of equations for 'belief systems' that are more general than probability theory; fuzzy set techniques, which provide a general set of equations for calculating 'degrees of set membership'; heuristic techniques, such as those based on endorsement theory; and several *ad hoc* techniques without formal theories supporting them.

A strategy for knowledge formalization

During the process of knowledge formalization, the problem being investigated is connected to its proposed solution by structuring the knowledge into a form acceptable to the system's processing environment. It is here that knowledge truly takes on its distinctive importance, for it is this knowledge which will control the generalization process. In the remaining discussion, a rule-based system environment will serve as a basis for reviewing the problem of knowledge formalization for digital generalization. Though other knowledge formalisms discussed above were shown to add certain benefits, production rules can provide a core resource of knowledge for each of those methods. Currently, two structures are appropriate for storing the knowledge in a rule-based system for conducting generalization operations: rules and parameter tables.

Rule structures

Production rules contain information about procedures and reasoning the inference engine processes. Rules provide the controlling means for determining the status of specific conditions which aid in making generalization decisions. The conditions that can be examined generally may be one of five types: (1) existence, which tests for the presence (or absence) of an item within the database; (2) scope, which tests for specific instances of some characteristic within a data set; (3) fact, a test for truth or fallacy of a fact; (4) value, which examines an entity's attributional values; and (5) relation, which addresses cartographic and topological relationships.

The action portion of a production rule typically affects three types of activities: (1) control, which alters the control logic that directs the search and reasoning techniques, or can adjust the currently active rule processing methodology to an alternate rule; (2) spatial transformations, which provide representational adjustments to the database entities in the spatial domain; and (3) attribute transformations, which modify the attribute(s) associated with a particular item under consideration.

Figure 1.3 illustrates a structural view of all possible condition–action combinations. A third dimension, the specificity of generalization applica-

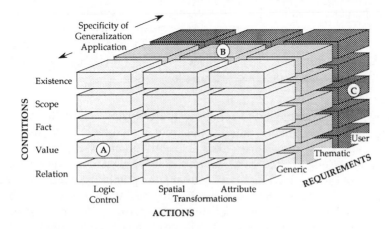

Fig. 1.3 Rule groups from production rules

15

tion, represents different levels of generalization 'severity' associated with generic mapping requirements, thematic requirements specific to a product type and scale of presentation, and requirements unique to a user. This severity component is akin to an object-oriented paradigm wherein hierarchies of interlinked objects organize the elements that form the final map product. That is, linked together are encapsulated pieces of data and procedures necessary for a particular level of representation. The linkage of objects, the communication between them, the capacity for co-ordinated processing among groups of objects, all become increasingly refined as knowledge is accumulated, and as the specificity of the generalization processing requirement is needed. Mark (1991, Ch. 6 this volume) discusses some rules for the generalization process using object-oriented representations.

According to the principles of this paradigm, the intersection of the three variables: conditions, actions, and requirements, would identify rule groups in an artificially intelligent digital generalization system. At a high level of abstraction, processing is analogous to a set of related objects which flow through a series of functional, executable programs, evaluating knowledge (rules) and elaborating the relationship between objects until the final set of objects is a sufficiently accurate representation of the final product. Rule groupings would be applied to various parts of the object hierarchy by the inference engine, consistent with the level of specificity required. The scope or subset of the objects upon which the inference engine would apply the rules is variable, and is specified by the inference engine which would alter or manipulate objects in the scope according to the rules.

Examples of production rules (for declarative knowledge) for three of the rule groups (labelled as A, B, and C in Fig. 1.3) are provided below.

A {Generic specificity}   IF <value>   THEN <logic>
    IF   product scale is ⩾ 50 000
    THEN   examine all rules relating to point features
B. {Thematic specificity}   IF <existence>   THEN <spatial>
    IF   railroad features are normally depicted
    THEN   displace all sidings at least 3.0 mm from railroads
C. {User specificity}   IF <fact>   THEN <attribute>
    IF   gas pipeline depth is unknown
    THEN   set navigation caution flag to true

Although these examples each represent the formalization of declarative knowledge of a quantifiable nature, this rule structure can be used to represent qualitative knowledge. Shea and McMaster (1989) outline six conditions to assess a need for generalization (congestion, coalescence, conflict, complication, inconsistency and imperceptibility conditions) and seven types of measures which can aid in judging the presence of such a condition (density, distribution, length and sinuosity, shape, distance, Gestalt, and abstract measures). Predicates and/or consequents in a production rule could represent these conditions and measures. It would not be unlikely to expect rules to state generalization heuristics such as those shown below.

{Generic specificity}　　　IF <relation>　　THEN <spatial>
　　IF　　　　　features coalesce at map scale
　　THEN　　　displace the feature of lesser importance
{Thematic specificity}　　IF <scope>　　　THEN <logic>
　　IF　　　　　distribution of buildings in urban areas is too congested
　　THEN　　　group buildings of similar nature to reduced complexity
{User specificity}　　　　IF <value>　　　THEN <attribute>
　　IF　　　　　methane level of a cave exceeds normal safety levels
　　THEN　　　identify the cave as unsafe for spelunking

Knowledge acquisition to support these types of production rules could be evolutionary, wherein straightforward, generic processing knowledge is represented first as generic processing rules, and the human being is subsequently relied on for control and heuristic knowledge in a more specific domain. Factual knowledge should be formalized first, with heuristic knowledge following as the declarative knowledge base is refined. In Chapter 2, McMaster (1991) begins to explore sources for this knowledge acquisition task. An interactive digital generalization system which allows individual application of generalization functions could provide the basis for both knowledge acquisition and knowledge refinement. Weibel and Buttenfield (1988) and Weibel (1991, Ch. 10 this volume) point out that this gradual refinement of knowledge supports an amplified intelligence concept, rather than one exhibiting AI. The ability to communicate the intuitive generalization heuristics during the process of knowledge acquisition will temper the development of a system exhibiting complete AI. Over time, the decisions made by human intervention could be modelled and structured as declarative and heuristic knowledge within each level of refinement, thereby adding greater knowledge to the thematic and user levels of the knowledge base.

Parameter table structures

Parameter tables provide the second means for representing a great deal of cartographic knowledge required for generalization decision-making. Developed as lists or tables of information, parameter tables store simple relationships for access by some relational referencing methodology during rule-based processing. In effect, parameter tables govern the way in which the production rules operate by binding specific generalization algorithms and parameters to specific generalization situations. Many tables can be easily populated by means of decomposing specifications for specific product types, and are an excellent technique for representing primarily declarative knowledge gained through an examination of general cartographic principles, processes, and procedures.

17

Parameter tables may be simple or complex in nature. To illustrate, a table of algorithm parameters for a linear simplification algorithm differentiated by feature type would be an excellent example. A simple case is a relational table which identifies differing algorithm tolerance values against feature type. On the other hand, Buttenfield (1991, Ch. 9 this volume) demonstrates the use of several parameters to create line geometry-based structure signatures as a means for controlling the linear generalization process. These more complex parameters may vary even within a single feature type.

In consonance with the production rules, a taxonomy of parameter tables would encompass most input, processing, and output variables necessary to control generalization operations. Parameter tables provide the means to formalize the data either assessed in the antecedent of a production rule, or modified as a result of an action taken on the consequent. Rather than duplicating specific assessments or directions in the rules themselves, all production rules would be indexed and referenced to the parameter tables. This process serves to aid in the standardization of nomenclature used in the rule-based system.

Using a relational database paradigm, wherein an indexing scheme maintains associations between parameter values sharing a common relation, several classes of parameter tables help organize the generalization knowledge. Figure 1.4 illustrates how several of these tables can relate to one another.

The **master parameter table** details the method of generalization appropriate for each object in a cartographic database, indexed to all input, processing, restriction, and output parameter tables. For example, all feature entities that can exist as both an area and point feature, which meet some specified input criteria, are candidates for generalization methods

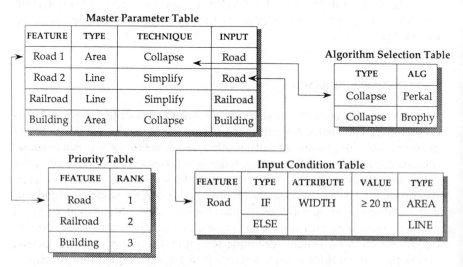

Fig. 1.4 Sample parameter tables

which collapse features. The appropriate restrictions on the newly created feature's representation would be identified by reference and an identification of the appropriate output attribution would be noted.

The **input condition table** provides an indication of whether or not the attribute evaluations, either explicitly recorded or calculable, meet the conditions specified in the predicate (antecedent) of a production rule. This table would contain an indexed set of all possible input attribute condition expressions consisting of attribute category (such as length), a relational operator (such as =, ≥, <, etc.), and a value with which to make the comparison.

The **algorithm selection table** identifies the specific algorithms available for the various generalization methods. For example, all features identified in the master parameter table that are subject to linear simplification, would have the specific algorithm(s) identified that that particular feature would be subjected to (such as Douglas, Lang, or Deveau), as well as a reference to the requisite parameters used to control the degree of application of each algorithm.

The **algorithm control table** consists of the parameters used to control each algorithm.

The **representation table** indexes all restrictive parameters associated with representing a feature on a product. For example, this table would tie together any portrayal priorities, displacement criteria, or rationality of depiction for all newly created and modified features.

The **priority table** specifies the order of generalization priority for determining the order in which feature symbols in a set of features will undergo generalization manipulations. Numerically ranked features set this priority.

The **displacement table** specifies the order of priority or hierarchy for determining which feature symbols in a set of features will be portrayed at ground truth and which will be displaced when graphic scale prohibits all being retained at ground truth.

The **rationality table** is used to specify if the interaction of newly created features with existing features makes sense from a cartographic and topological viewpoint. This table would specify all possible contingent relationships between point, line, and area features.

The **output condition table** provides an adjustment to attribute values based upon the outcome of some generalization operation. This table would contain an indexed set of all possible output attribute condition expressions consisting of attribute category, an assignment operator (=), and a value with which to make the comparison.

Summary and recommendations

This chapter has addressed several implementation issues confronting the development of a rule-based digital generalization system. The concept embodies methodologies which reflect some aspects of human intelligence, although the goal is not to produce a complete mimic of the human generalization process. Previous attempts at automation have been limited by a lack of an understanding of the formal logic of the generalization process. An overview of expert systems demonstrated the integral nature that knowledge plays in constructing an artificially intelligent system, and the degree of knowledge resident within the system has a direct influence on the power provided by the system.

Design considerations for an artificially intelligent digital generalization environment might be implemented using a rule-based system approach. A proposed strategy for knowledge formalization within such a system requires that experts' knowledge can be conveniently organized by production rules, coupled with the appropriate set of parameter tables. Knowledge organization must support updates or corrections easily as new information becomes available. A taxonomy of rule group and parameter table structures was presented as a method of organizing information necessary for controlling the generalization process. These structures can be incrementally developed and populated as an understanding of the digital generalization process is obtained.

2

Conceptual frameworks for geographical knowledge

Robert B. McMaster

Introduction

For the development of fully operational automated software for cartographic generalization, three complex problems must be solved: (1) a
formal, comprehensive conceptual framework for digital generalization must
be agreed upon; (2) the specific procedures of the process, or generalization
operators, must be designed, coded, and tested; and (3) cartographic
knowledge must be culled from expert sources (maps) and individuals and
coded into 'rules'.

This chapter explores each research area and illustrates the relationship
between the three. First, a series of generalization models are reviewed and
one, developed by Brassel and Weibel, is identified as the best for
implementing an expert system. Next, the development of generalization
operators is traced and a comprehensive listing is provided. These
generalization operators are placed within a logical framework which fits
well within the Brassel and Weibel model. Lastly, an existing rule base,
developed by the US Defense Mapping Agency (DMA), is evaluated for its
potential use in expert systems. It appears that many of the rules may
logically be structured within the 'process library' of a generalization system.
The process library provides the structural framework for both rules and
procedures within an expert system.

Models of generalization

Throughout the 1970s and 1980s a multitude of generalization models have
been developed, both in the European and American literature. Some of
the models have addressed specific components of the generalization

process, such as the manipulation of attribute information, while other models have been more comprehensive. A review of five such models illustrates the breadth of approaches developed to describe the generalization process.

Ratajski model

In the mid 1970s the Polish cartographer Lech Ratajski presented one of the first formal models of generalization. In the work, entitled 'Phénomènes des points de généralisation'. Ratajski (1967) identified two fundamental types of generalization processes: quantitative and qualitative (Fig. 2.1). Quantitative generalization involves a gradual reduction in map content which is dependent on scale change, while qualitative generalization results from the transformation of elementary forms of symbolization to more abstract forms. An important component of Ratajski's framework is the **generalization point**, or the point in scale reduction at which the map capacity is decreased to the level where a change in the cartographic method of representation is necessary (e.g. Fig. 2.1, A to B). This changing map capacity may be represented as the apex of a triangle, with the base of the triangle depicting maximum capacity and the apex depicting the limit (minimum capacity). A horizontal slice through the triangle, positioned parallel to the base, represents a given level of generalization. At the apex

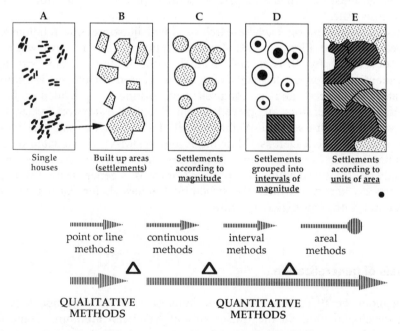

Fig. 2.1 Ratajski model of generalization (after Ratajski 1967)

of the triangle, of course, minimum capacity is reached and a new cartographic method of symbolization must be applied.

For instance, as illustrated on Fig. 2.1(B), individual homes must be replaced with new symbolization for the settlement as a whole, which is considered a point or line method. With continued generalization, the built-up areas are converted to settlements according to magnitude (termed a continuous method), settlements according to intervals of magnitude (interval method), and into settlements according to enumeration units (areal method). Figure 2.1 graphically illustrates these transformations (A–E) and also identifies the differences between qualitative and quantitative methods.

Morrison model

Using the structure defined by Robinson and Sale (1969), Joel Morrison established a formal model of generalization based on set theory (Morrison 1974). Figure 2.2 represents the Morrison model, in which the generalization process consists of the transformation of the sensory elements of the cartographer's reality (SCR) to the physical elements of the map (PM).

This transformation was labelled f_1, which is illustrated on Fig. 2.2. A second transformation, g_1, relates the physical elements of the map, PM, to the sensory elements of the map reader's reality, or SRR (Morrison 1974: 117). Morrison then viewed each of the generalization processes (simplification, classification, induction, and symbolization) in terms of their two probable transformation characteristics (one-to-one and onto). A discussion of the specific mathematical properties of one-to-one and onto, as applied to the process of cartographic generalization, may be found in the original paper by Morrison. For instance, selection, a preprocessing step to generalization, is considered to be both one-to-one and onto, because each element on the original map has one corresponding element on the generalization and the process is reversible. The generalization process of classification, alternatively, is applied after selection and involves such

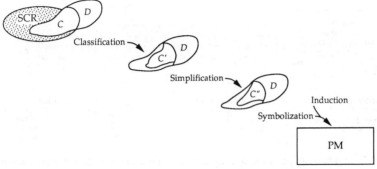

Fig. 2.2 Morrison model of generalization (after Morrison 1974)

activities as grouping rivers or towns, or combining land use categories. Classification, in terms of the transformational characteristics, $f_{I_{ij}}$: C–C′, is one-to-one, but cannot be onto, while the inverse ($f_{I_{ii}}^{-1}$: C–C′) is not one-to-one, but is onto. The other elements of simplification, induction, and symbolization are likewise described.

Nickerson and Freeman model

A model of the generalization process, specifically designed for an expert systems approach, was presented by Nickerson and Freeman (1986). As illustrated on Fig. 2.3, their model consists of five tasks: (1) the four distinct feature modification operations; (2) symbol scaling; (3) feature relocation and symbol placement; (4) scale reduction; and (5) name placement. The problem of map generalization, as described by the authors, is one of effectively converting the source map, of scale **1 : m**, symbol size **a**, and area **w** ∗ **h**. The generalization operators, or feature modification operations, of this model will be detailed later in the chapter.

After the four feature modification operators have been applied at the original scale (feature deletion, simplification, combination, and type conversion), the symbols in the (modified) source map are scaled to increase their size by the factor **k**, where **k** is greater than unity. Conceptually, as illustrated on Fig. 2.3, this process yields an intermediate map of scale **1 : m**, symbol size **ka**, and area **w** ∗ **h**. It is at this intermediate map scale (**1 : m**) that features are relocated and symbols replaced due to overlap and interference. The second scale reduction (based on the intermediate map) produces the target map, of scale **1 : km**, symbol size **a**, and area (**w/k** ∗ **h/k**). After the final scale reduction, a last generalization task involves name placement.

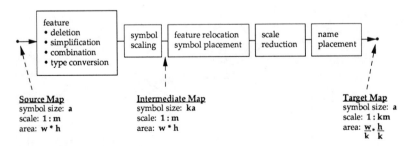

Fig. 2.3 Nickerson and Freeman model of generalization (after Nickerson and Freeman 1986)

McMaster and Shea model

A model developed by McMaster and Shea (McMaster and Shea 1988; Shea and McMaster 1989) identifies three considerations for comprehensive generalization, including: (1) intrinsic objectives, or why we generalize; (2) situation assessment, or when we generalize; and (3) spatial and attribute transformations, or how we generalize. In expanding on the nature of why we generalize, three distinct types of intrinsic objectives were identified: philosophical objectives, application objectives, and computational objectives. Fundamental to the development of the model is the establishment of six philosophical principles of generalization, including: (1) reducing complexity; (2) maintaining spatial accuracy; (3) maintaining attribute accuracy; (4) maintaining aesthetic quality; (5) maintaining a logical hierarchy; and (6) consistently applying generalization rules.

Reduction of graphic complexity is cited as the most important principle overall. Although cartographers have never developed an adequate measure of graphic complexity – in a visual sense – a basic definition is the measure of the interaction of the various graphic elements within a map or, specifically, the number and/or diversity of these elements within a given area (McMaster and Shea 1988). Furthermore, identifying, analyzing, and defining appropriate levels of complexity is perhaps the most difficult problem in digital generalization, for it requires that many of the spatial and attribute transformations be applied either iteratively or simultaneously (McMaster and Shea 1988).

Most of the generalization operators that have been developed, such as simplification, smoothing, and displacement, of course, relate to the detection of and adjustment for complexity. For example, simplification algorithms determine where coordinate density has exceeded a given tolerance level and weed unnecessary data based on some geometric/ mathematical criterion. Smoothing algorithms attempt to 'plane away' overly complex detail along a linear feature. Displacement routines detect coalescence of graphic features and shift them apart in order to decrease complexity. Understanding graphic complexity, then, is viewed as a crucial component of digital generalization.

A modified definition of digital generalization is provided in the model, which states that generalization, in a digital environment, requires the application of both spatial and attribute transformations in order to maintain clarity, with appropriate content, at a given scale, for a chosen map purpose and intended audience (McMaster and Shea 1988). The spatial and attribute transformations, or generalization operators, will be discussed later in this chapter.

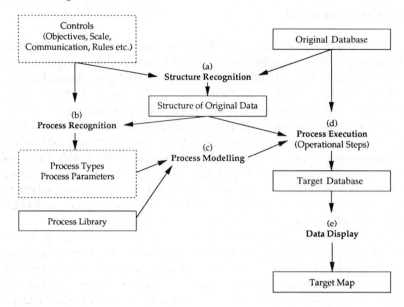

Fig. 2.4 Brassel and Weibel model of generalization (after Brassel and Weibel 1988)

Brassel and Weibel model

The model of digital generalization best suited for the integration of expert systems was developed by Swiss cartographers (Brassel and Weibel 1988) (Fig. 2.4). The model identifies five processes of generalization in a digital environment, including structure recognition, process recognition, process modelling, process execution, and data display.

Structure recognition is the activity identifying specific cartographic objects, or aggregates of objects, as well as the spatial relations and measures of importance. Structure recognition is controlled by original database quality, target map scale, and communication rules. Process recognition identifies the exact generalization operators to be invoked and involves both data modification and parameter selection. Process recognition specifically determines what is to be done with the original database, what types of conflicts have to be identified and resolved, and which types of objects and structures are to be carried in the target database. This is followed by process modelling, which compiles rules and procedures to apply from the process library. It is at this stage where the algorithms, such as simplification, smoothing, and displacement, are applied using both rules and parameters. Actual generalization takes place during process execution, where rules and procedures are applied to the original database in order to create the generalized output. A last process is data display.

Intrinsic to Brassel and Weibel's framework is the component termed **process library**, which contains the rules and procedures for generalization.

In building an expert system for automated map generalization, the development of such a process library will entail critical decisions involving what generalization operators are necessary and how might they be sequenced, what knowledge must be captured within the system (rules) and how might it be structured, and what parameters and tolerances are required for logical implementation of rules and operators. Thus, the three components of a comprehensive process library include operators, knowledge, or rules for generalization, and tolerance values. The remainder of this chapter will focus on two crucial components in the development of a rule base: the identification of unique processes or what will be called the generalization operators and the development of the related geographical knowledge.

The development of generalization operators

Various researchers have attempted to isolate the basic operators that are utilized in numerical generalization. In many ways, such operators are similar to those in the GIS software, for example, the Map Analysis Package, or MAP (Berry 1987; Tomlin 1983). In the case of Tomlin's work, however, these are known as spatial operators. By 1978, one entire chapter of Arthur Robinson's *Elements of Cartography* had been devoted to the topic of generalization, where the four elements used by Morrison in his model – simplification, classification, symbolization, and induction – were fully explained. Simplification was defined as the determination of the important characteristics of the data, the retention and possible exaggeration of these important characteristics, and the elimination of unwanted detail. Classification was identified as the ordering or scaling and grouping of data, while symbolization was defined as the process of graphically encoding these scaled and/or grouped characteristics.

Brassel generalization operators

Brassel (1985) developed the idea of a function and object concept, where generalization functions, such as expand, displace, eliminate, and smooth may be applied to an individual object or combined objects or features. Brassel provided categories of objects, including point, line, composite, area, and volume features. Additionally, an extensive list of operators is provided and is divided into point, line, area, volume, and 'text' operators (Table 2.1).

For instance, in identifying techniques that may be applied to line features, Brassel identifies such activities as: (1) expand/shrink/select/ eliminate; (2) reduction of sinuosity; (3) change topology in linear networks; (4) displace; and (5) classify. Although Brassel's list of operators does refer

Table 2.1 Brassel's list of generalization operations, after Brassel (1985) (text and space-object operations not included)

Point features	Line features
Select point features subject to certain criteria	Expand/shrink/select/eliminate
Eliminate point features subject to certain criteria	Reduction in sinuosity
Identify clusters of point features and create a smooth polygonal outline	Change topology in linear networks
Expand	Displace
Displace	Clear
Classify	Classify
Display	Overlay line networks (intersect, merge, etc.)
	Convert raster to vector
	Reduce
	Display
Area features	**Volume features**
Select	Select, eliminate
Eliminate	Classify
Expand/shrink	Overlay (logical and arithmetic
Reduce sinuosity	operations)
Change topology	Find gradients/slopes
Classify	Convert to grid Triangulated Irregular Network (TIN)/contour
Displace	Recognize local minima/maxima
Clear	Recognize specific shapes
Overlay (logical and arithmetic operations	Eliminate specific shapes
Convert raster to vector	Recognize specific objects
Replace area feature by point/line feature	Replace volume features by area/line/ point feature
Display	Find ridge/course line skeletons
	Modify ridge/course lines
	Smooth surface
	Display

to volume features, little substantive work has been completed in this area. Since most of the effort thus far has focused on the generalization of data in vector format, ignored have been the problems of the generalization of topographic surfaces based on digital terrain models, or truly 'volumetric' features. A recent conference proceedings (Brassel 1990) on digital generalization does briefly address topographic generalization, although the emphasis is not on digital evaluation models (DEMs), but on traditional topographic maps. It is hoped that such work will instigate further detailed studies in topographic generalization.

Buttenfield generalization operators

Buttenfield, in a paper reviewing the various theories of cartographic lines, isolated a set of manipulations normally applied in line generalization (Buttenfield 1985). In addition to Robinson and Sale's (1969) conventions of simplification, selection, classification, and induction, Buttenfield added the concept of enhancement. Enhancement is defined as introduction of detail into a (line) feature to augment or emphasize particular detail or characteristics. It may involve any introduction of detail, for example, such as fractal enhancement.

DeLucia and Black generalization operators

DeLucia and Black (1987), in describing the design of a comprehensive generalization system for Intergraph, listed five basic functional processes for map generalization, including simplification, agglomeration, aggregation, feature collapse, and distribution refinement. Simplification was defined as the elimination of unwanted detail in line features and areal outlines. Agglomeration is the merging of two or more area features into a new feature, while aggregation was defined as the merging of three or more point features into a new area feature. The more complex operation of feature collapse involved the reduction of the spatial extent of a feature to the degree that it assumes a different cartographic form, such as the collapse of an area to a single point. Lastly, distribution refinement is the process of retaining only a representative pattern of features in order to 'typify' a feature at reduced scale.

Beard generalization operators

Yet another set of functions, or generalization operators, were presented by Beard (1987) for application to categorical data coverages, such as soil maps or land use maps. These five functions included: select, aggregate, reduce, collapse, and coarsen. The select function was defined as an operation to retain or eliminate entire classes of features. Selection criteria include both geometric or attribute data. The function aggregate condenses the attribute information by reducing the number of classes. In the reclassification process the geometry, also, is manipulated. Reduce removes coordinate information along linear features. Collapse, in which areas are compressed to lines or points, necessarily requires a change in dimension. A last function, coarsen, removes or modifies features by analysing clusters of points which fall within some distance epsilon of each other (Beard 1987). Beard also realized that each of the functions can be invoked in different combinations and order.

Nickerson and Freeman generalization operators

In the work on expert systems and generalization by Nickerson and Freeman (1986), five separate tasks, or feature modification operators, were identified. These included: feature deletion, feature simplification, combination, interference detection, and feature displacement. Feature deletion is the process of either eliminating whole classes of features from the entire map (e.g. urban areas), or weeding features from high-density regions on the map. Feature simplification involves the elimination of coordinate pairs along a polyline. Combination blends together linear features, such as the two edges of a riverbank or multiple features of a railroad yard. Lastly, interference detection and feature displacement are the processes of (1) determining if two features collide and (2) the subsequent adjustment of one feature in order to prevent graphic coalescence. Possible mathematical solutions are provided for each of these operations.

McMaster and Monmonier generalization operators

In a recent set of papers, McMaster and Monmonier establish a formal set of generalization operators (McMaster 1989b; McMaster and Monmonier 1989) (Fig. 2.5). Although this diagram was originally labelled 'A model of digital cartographic generalization', a better description might be 'A framework for generalization operators'. The framework initially isolates geographical and attribute data as the two basic forms encoded in digital cartography. Attribute data, such as census tract information, or soil categories, must receive different treatment from the geographical form, such as a polygon arc, in the generalization process. Attribute generalization includes the processes of both classification and symbolization. The shaded box labelled SELECTION PROCESS represents a preprocessing step to actual generalization. Before the objects or attributes are manipulated by the generalization operators, decisions are made as to what will be retained or eliminated. Selection, then, may be defined as a series of binary decisions. The division of generalization operators, positioned below the selection process, into raster and vector is based on the logical organization of geographical space into two models.

The vector data model treats a map as a geometric line structure whereas a raster data model focuses on maps as images in which the logical element is the cell, or pixel. These two approaches to organizing space are also termed location-based and object-based (Peuquet 1988b). Most of the recent activity in numerical generalization has focused on methods for object-based generalization, although a paper detailing the four major categories of operators for 'location-based' methods (structural, numerical, numerical categorization, and categorical), has recently been published (McMaster and Monmonier 1989). There is a sound justification for such a division of generalization operators, based on the development of recent

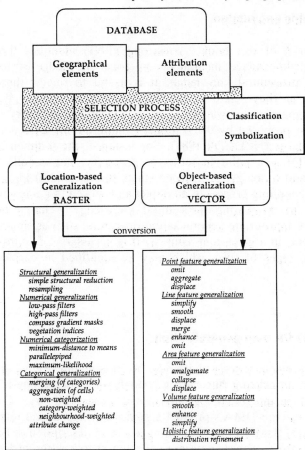

Fig. 2.5 A framework for digital generalization operators (after McMaster and Monmonier 1989)

GIS software, such as ARC/INFO and System 9 (object-based), and MAP and IDRISI (location-based). The most commonly applied vector-based generalization operators are categorized under the headings: point-feature generalization, line-feature generalization, area-feature generalization, volume-feature generalization, and holistic-feature generalization. Any comprehensive generalization package should have available many operators, although which specific operators are provided is dependent on user requirements.

Cartographic knowledge

In Chapter 5 of this book, Armstrong (1991) identifies three types of cartographic knowledge, including geometrical knowledge, structural knowledge, and procedural knowledge. It is useful to review these classes of knowledge as they relate to the development of specific 'generalization rules'. Geometrical knowledge refers to the actual geometry or topology, such as that proposed by the National Committee for Digital Cartographic Data Standards (NCDCDS 1988). For instance, the 0-dimensional phenomenon (0-D) is a point in terms of geometry, but a node in terms of geometry and topology. Armstrong states that structural knowledge arises from the generating process of an object and is used to guide generalization. According to Armstrong, procedural knowledge is based to guide the selection of appropriate generalization operators, such as simplification and displacement, in a given map context. It is necessary, of course, to look at the precise types of rules which may be identified in each of the three classes.

A rule base for map generalization

Using the previously described model by Brassel and Weibel and classes of knowledge, an existing rule base was analysed to determine if the existing rules could be integrated logically into a 'process library'. The rule base evaluated was the DMA's product specifications for digital feature analysis data (DFAD) (DMA 1986). Following is a brief description of the rule base.

Digital feature analysis data are collected at two different levels, termed level 1 and level 2. Level 1 is defined as 'a generalized description and portrayal of planimetric features' such as vegetation, soils, water, and cultural features, that are useful in relatively large planning and high-elevation flight simulations (DMA 1986: 1). The density of detail approximates that of medium-scale (i.e. 1 : 200 000–1 : 250 000) cartographic products. Level 2, on the other hand, is described as a highly detailed description and portrayal of planimetric features and 'is intended to cover small areas of interest and has smaller minimum size requirements for portrayal of planimetric features' (DMA 1986: 1). Level 2 density approximates that of 1 : 50 000 scale maps.

Three types of features – point, line, and area – are used to describe both natural and man-made objects, or entities. The basis for each feature is the 'delta pair'. The DMA defines the delta pair as follows: 'Each coordinate in a feature is represented by difference measurements, called delta values, in integer tenths of seconds which, when added to the value of the manuscript origin, describes coordinate location' (DMA 1986: I-1). Areal entities are depicted by a series of delta pairs that completely enclose an object or group of related objects and are defined by specific attributes. Linear entities are depicted by a series of delta pairs where length is more significant than width and are also defined by specific attributes. Point entities are depicted with a single delta pair which identifies the centre of the object.

Each of the features in the database is also identified with a unique number-code identifier referred to as a feature analysis code (FAC). The document also provides a set of general rules for each type of feature, where, for example, a rule for an areal feature is 'All areal features contain an implicit delta pair numbering system which starts with delta pair one and progresses continually and sequentially until the feature area has been completely delimited. Consecutive delta pairs do not have the same geographic value' (DMA 1986: 8). Most importantly, it appears as though the rules developed by the DMA do not relate to scale change modification, but only for display at the specified portrayal scales for each level.

Table 2.2 DMA digital feature analysis data (DFAD)
Section 452. Railroads, railroad yards, and sidings

452. RAILROADS, RAILROAD YARDS AND SIDINGS
  A. General
    1. All railroad routes will be portrayed.
    2. Railroads will be portrayed as continuous linear features throughout their route
      a. When railroads are in a cutting or supported by a linear feature such as an embankment, bridge, or causeway, the supporting feature will be portrayed as a separate linear feature occupying the same alignment as the railroad. The uppermost feature (the railroad) will be assigned the higher FAC number. Height of the railroad is equal to the height of the elevated supporting feature or depth of the cutting.
      b. When the supporting bridge is portrayed as a point feature falling on the same alignment as the railroad, the supporting bridge will be assigned the higher FAC number.
      c. Height is standardized as zero for railroads not associated with a supporting feature or cutting.
    3. The width of each set of tracks is standardized as two meters.
    4. The directivity is coded as bi-directional, and the standard mapping category (SMC) is standardized as '2'.
    5. Multiple track railroads will be portrayed as a single linear feature The width will be the number of tracks multiplied by the standardized width per track.
    6. When a railroad route passes through a railroad yard, the yard will be shown as a separate linear feature(s).
  B. Railroad yards and sidings which are a minimum of 300 meters in length and five tracks wide (including the through tracks) will be portrayed as linear features representing the track pattern. The center of the yards, spurs, or sidings track pattern will be portrayed as the linear feature. The width of each track pattern will represent the total number of tracks in the pattern multiplied by the standarized width per track.

Table 2.3 DMA digital feature analysis data (DFAD)
Section 453. Roads

---

453. ROADS
    A. General
        1. Roads will be portrayed as continuous linear features throughout their route.
            a. When the roads are in a cutting or supported by a linear feature such as a bridge, embankment or causeway, the supporting feature or cutting will be portrayed as a separate linear feature occupying the same alignment as the road. The uppermost feature (the road) will be assigned the higher FAC number. Height of the road is equal to the height of the supporting feature or the depth of the cutting.
            b. When the supporting structure (bridge) is portrayed as a point feature falling on the same alignment as the road, the supporting structure (bridge) will be assigned the higher FAC number.
            c. Height is standardized as zero for roads not associated with a supporting feature or cutting.
        2. The width of each lane is standardized as 4 meters, excluding non-hard-surfaced shoulders.
        3. Roads will be portrayed as a single linear feature. The width will be the predominant number of lanes times the standardized width per lane, excluding non-hard-surfaced shoulders.
        4. The directivity is coded as omni-directional.
    B. Dual Highway with Median – A dual lane highway is defined as an all weather, primary road, four or more lanes wide, separated by a median, barrier, or parkway.
        1. All dual lane highways with medians will be portrayed (including those within urban areas).
        2. The SMC shall be '9' or '14'.
        3. When a separation (median) greater than 100 meters between lanes occurs, the lanes will be portrayed as separate features.
    C. Primary roads. A primary road is defined as an All Weather Hard Surface or All Weather Loose or Light Surface Road.
        1. All primary roads which form the principal road transportation network of a region will be portrayed.
        2. Primary roads are not portrayed within urban areas (SMC 1–4), unless they extend beyond the outer boundaries of the urban area and are part of the selected road network of the region.
        3. SMC shall be '5', '9' or '14'.
    D. Secondary Roads. A secondary road is defined as a Fair Weather Loose or Light Surface Road, a Cart Track or Trail.
        1. Secondary roads will only be portrayed in areas of sparse culture in which they constitute the *only* transportation network. Portrayal will be limited to those having a definite strategic purpose such as connecting outposts with one another, connecting widely separated primary roads or emanating from a road termination to definite objectives which are portrayed.
        2. SMC shall be '5'.

The DMA rulebase

Table 2.2 presents the DMA rules related to one class of line features, specifically 'Railroads, railroad yards and sidings' (category 452). Using this table, it is possible to translate some of these generalized descriptions into distinct categories, based on Fig. 2.5. For instance, rule 1 relates to selection: 'All railroad routes must be portrayed'. Rule 5, 'Multiple track railroads will be portrayed as a single linear feature' relates to line amalgamation, which involves a geometric manipulation. Rule 6 relates to the process of typification.

It appears as though four different types of rules may be identified from the rule base: general rules, selection rules, hierarchical rules, and geometric rules. On p. 34 is a set of rules for road features (Table 2.3). For instance, rule 1, which states that 'roads will be portrayed as continuous linear features throughout their route', is a general rule, implying that linear features should not be broken. Also note the definition provided for rule C, 'A primary road is defined as an All Weather Hard Surface or All Weather Loose or Light Surface Road.' This also represents a general rule. Yet another example of a general rule – not listed in the table – is (under 'Road interchanges') 'Height for linear bridges and for point bridges which are

Table 2.4 DMA Digital feature analysis data (DFAD)
Section 482. Isolated structures

---

482. ISOLATED STRUCTURES
 A. Features such as tanks, silos, exposed wrecks and buildings are significant, particularly in areas of sparse culture.
 B. Only SMC 1, 2, or 3 structures will be selected for portrayal.
 C. Isolated structures are defined as those structures which are located more than 150 meters Level 1 or 30 meters Level 2 from an SMC 1–4 areal outline. SMC 1 and 2 isolated features are exempt from this limitation.
 D. If structures are too densely spaced to portray as individual features, a representative pattern will be selected for portrayal.
 E. The height for mobile homes is standardized as 4 meters.
 F. Buildings that are too small for individual portrayal should be considered (analyzed) as a group only if they are densely spaced. The open space between buildings cannot comprise more than 20% of the building grouping.
 G. These features may also be shown as linear features when the length of the feature is equal to or greater than 150 meters Level 1 or 30 meters Level 2.
 H. Strings of buildings may be portrayed as a linear feature if the combined length of the buildings is equal to or greater than 150 meters Level 1 or 30 meters Level 2 and equals 80% of the line length, and all other minimum requirements are met. The linear feature must reflect the actual position of features at either end.
 I. Isolated farmhouses or groups of agricultural buildings of SMC 3 material that meet the point feature selection criteria may be described with a standardized description of length and width = 44 meters Level 1, or 20 meters Level 2; height = 8 meters; orientation = 360 degrees; feature identification code (FID) = 430.

---

elevated above the local terrain will be the highest portion of the bridge deck above the water/terrain. The SMC will reflect the predominant surface material of the entire bridge structure.' Rule 1a states, 'the uppermost feature (the road) will be assigned the higher FAC number'. This rule is hierarchical and implies that other features should be broken for the road (FAC numbers identify objects and prioritize their portrayal). Rule B1, 'All dual lane highways with medians will be portrayed' is an example of a selection rule.

A good example of both selection and geometric rules may be found under the 'Isolated structures' category (Table 2.4). Initially, a general rule is provided: 'Features such as tanks, silos, exposed wrecks and buildings are significant, particularly in areas of sparse culture.' This is followed by two selection rules: B 'Only SMC1, 2, or 3 structures will be selected for portrayal. C Isolated structures are defined as those structures which are located more than 150 meters Level 1 or 30 meters Level 2 from an SMC 1–4 areal outline. SMC 1 and 2 isolated features are exempt from this limitation.' Next, a geometric rule states: D 'If structures are too densely spaced to portray as individual features, a representative pattern will be selected for portrayal.' Thus it appears that a rule base structure would require that categories of rules be established, and applied sequentially.

After a careful examination of many examples of DMA rules, it is apparent that certain 'categories' of both knowledge and, more specifically, rules, have been captured. General types of rules that have been coded – such as 'Features such as tanks, silos, exposed wrecks and buildings are significant, particularly in areas of sparse culture' – represent structural cartographic knowledge, since they 'bring geographical expertise that ordinarily resides with the cartographer' (Armstrong 1991: 69). Likewise, selection rules such as 'Only SMC1, 2, or 3 structures will be selected for portrayal' are structural. On the other hand, both hierarchical rules and geometric rules represent procedural knowledge, since they 'are used to select appropriate operators for performing generalization' (Armstrong 1991: 69). For instance, 'When the supporting bridge is portrayed as a point feature falling on the same alignment as the railroad, the supporting bridge will be assigned the higher FAC number' is a hierarchical rule and would result in a displacement of features away from the bridge. Likewise, the geometric rule 'If structures are too densely spaced to portray as individual features, a representative pattern will be selected for portrayal' would invoke the generalization operator typification. The geometric knowledge encodes the individual feature descriptions, but does not directly relate to absolute rules.

A framework for the process library

In order to view the interaction of the generalization model, generalization operators, and cartographic knowledge, a modified diagram is provided (Fig. 2.6) which focuses on the structure recognition, process modelling, and process library components of Brassel and Weibel's model. The initial process is structure recognition after which a selection filter, based on selection rules, is applied. The selection rules are invoked by the controls of the process – scale and map purpose. Next, as identified by Shea and McMaster (1989), a set of conditions (congestion, coalescence, conflict, complication, and inconsistency) are tested for with a variety of measures, such as density of multiple features, length and sinuosity of single features, and the Gestalt, or perceptual characteristic of the information.

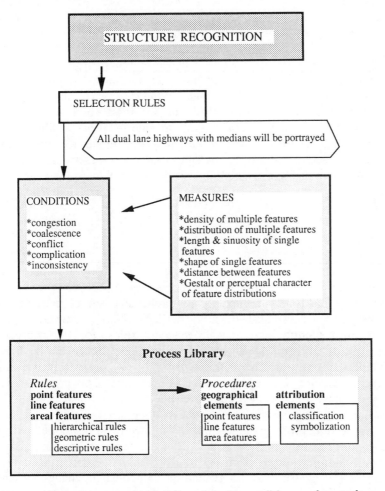

Fig. 2.6 Modified Brassel and Weibel model with conditions and controls

Process modelling, the next activity, requires a detailed process library, with both rules and procedures. Here, a logical design separates the geometric operators and types of data (point, line, and area). For each data type, different categories of rules may be invoked, including descriptive rules, hierarchical rules, and geometric rules. After rule processing the appropriate generalization operator – simplification, smoothing, displacement – is applied to the resulting data. The entire process may, of course, be iterative in nature, requiring several applications of both rules and operators.

The bottom box on Fig. 2.6 represents the process library. The structure for this library, as discussed previously, is organized into both a rules and procedures component. For each of the feature types (point, line, and area), hierarchical, geometric, and descriptive rules must be established and applied to features. The procedures, traditionally called generalization operators, are applied to the features both before and after application of rules. For instance, it may be that, due to the application of a hierarchical rule, a road is first displaced away from a railroad and is subsequently simplified based on a geometric rule. The sequencing and iteration of rules and procedures are extremely complex and researchers are still many years away from providing specific guidelines. Furthermore, a great deal of additional research is necessary in designing the specific components of, and structure for, the process library. Undoubtedly, researchers will eventually discover that individual – application specific – process libraries are necessary. Thus a process library designed for digital elevation model generalization would have different rules and operators from one implemented for generalizing thematic land use/land cover maps.

Summary

It is clear that for the successful implementation of rule bases, and ultimately for fully operational expert systems, appropriate models of generalization must be designed, a comprehensive set of generalization operators must be established, and, perhaps most importantly, cartographers must identify and logically encode 'rules' for the process. It appears that an excellent starting-point is the model developed by Brassel and Weibel (1988), where the concept of a process library, which contains rules and procedures for generalization, is presented. This process library, of course, must thus be divided into one section which contains the generalization operators and another which organizes the rules.

Many procedures for digital generalization have been developed over the past 20 years. For example, the problems associated with line simplification have been addressed from a variety of research paradigms, including the development of algorithms, the evaluation of geometric change, resulting from simplification, and the automatic modification of tolerance values in order to obtain consistent results (Buttenfield and Mark 1991). Substantial work has also been completed on displacement and smoothing. One significant gap in the current literature is the lack of empirical work on the sequencing and interaction of operators. Formal rules for digital map generalization are almost non-existent.

By evaluating an existing rule base used for digital generalization, the DMA's product specifications for DFAD, several categories of rules have been identified, including general rules, selection rules, hierarchical rules, and geometric rules. Interestingly, these rules may be further subdivided into those designed to modify point, line, and areal objects or features. Thus a rudimentary organization of the process library may be established, with the point, line, and area generalization operators linked to the same types of rules. Although actual rules will change substantially from one database and application to the next, such analysis allows the structure of the process library to be determined.

Acknowledgement

Robert B. McMaster completed this research while on the faculty of Syracuse University.

3

Knowledge engineering for generalization

Bradford G. Nickerson

Introduction

The Canadian National Topographic Series of maps uses 12 922 1 : 50 000 scale map sheets defined by geographical coordinate boundaries as its base (see Fig. 3.1). From these base maps are derived (using cartographic generalization) a series of 1 : 250 000 scape map sheets. Each 1 : 250 000 map sheet is of size 1° latitude by 2° longitude south of 68°, 1° latitude by 4° longitude north of 68° and south of 80°, and 1° latitude by 8° longitude north of 80°. There are 16 1 : 50 000 map sheets covering the same area as one 1 : 250 000 map sheet. The generalization is currently carried out by experienced cartographers at the Canada Centre for Mapping (CCM) in Ottawa, or by private companies under contract. This ongoing cartographic generalization activity presents a valuable opportunity to study currently used manual cartographic generalization methods, with a view to codifying them. From the expert system development point of view, this can be considered a knowledge engineering task, although in some respects it is much more difficult.

Previous attempts at automated generalization

Ever since computers have been used to process map data, there have been attempts to generalize map data automatically. For example, Tobler (1964), Ramer (1972), Douglas and Peucker (1973), and Dettori and Falcidieno (1982) all report on methods for polyline simplification (i.e. reducing the number of points representing a linear feature while retaining its characteristic shape). Automated generalization of complete cartographic features has also been attempted, with both vector format data (e.g.

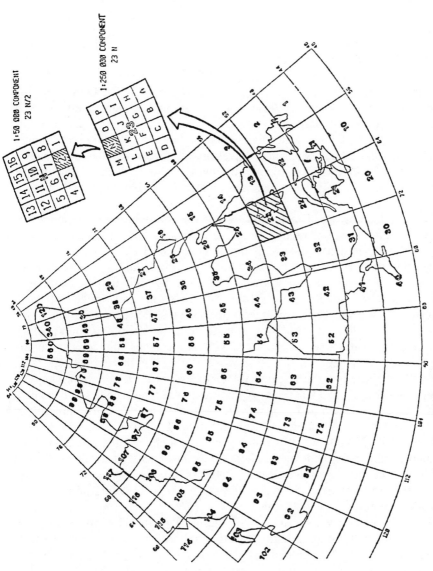

Fig. 3.1 Conceptual view of Canada's National Topographic Series of maps

Monmonier 1989b; Nickerson 1988a; Richardson 1988; Beard 1987; Chrisman 1983; Lichtner 1979) and with raster format data (e.g. O'Brien 1988; Monmonier 1987; Wilson 1981; Christ 1978). These attempts have demonstrated the difficulty of performing automated cartographic generalization.

The 1 : 250 000 pilot project

An interesting experiment was performed at the CCM In Ottawa during 1986 and 1987 (Brown 1987b). This experiment is important since it represents a serious attempt to use an algorithmic approach to generalize automatically 1 : 50 000 scale digital map data to 1 : 250 000 scale digital map data files. The experiment investigated three main areas. Feature selection was investigated by identifying the different requirements for 1 : 50 000 and 1 : 250 000 maps; some features can be eliminated outright (e.g. contours not required at the 1 : 250 000 scale are dropped). Data volume reduction was investigated by polyline simplification methods. File merging and edge matching were investigated by forming a single 1 : 250 000 scale digital data file from 16 generalized 1 : 50 000 scale digital data files. A fourth area of investigation was not pursued, as the basic rules of generalization were not known. This area focused upon cartographic quality control, and feature displacement, deletion, and symbolization to realize a cartographically acceptable 1 : 250 000 scale representation.

A feature selection table was defined to show (approximately) which features are eliminated on the 1 : 250 000 scale maps, but appear on the 1 : 50 000 scale map. Table 3.1 shows a portion of this table. The table shows that there is a third option for features besides simply being kept or dropped at the 1 : 250 000 scale. A feature can also be recoded to become a different type of data at the new scale. Table 3.1 shows an example of an airport runway represented as a line feature at 1 : 50 000 becoming a point feature at 1 : 250 000. Other examples include a change of feature type; for example, 'dry river bed' at 1 : 50 000 becomes 'sand' at 1 : 250 000, and 'seawall' becomes 'shorelines' at 1 : 250 000. Of the 394 feature types on 1 : 50 000 scale maps, 154 were dropped, 200 were kept, and 40 were recoded for the 1 : 250 000 scale.

For the remaining experiments of the 1 : 250 000 pilot project, hypsographic and hydrographic data from four adjoining 1 : 50 000 map sheets were used. Table 3.2 summarizes the amount of data reduction achieved using the feature extraction process mentioned above and a polyline simplification algorithm. On average, this represents a saving of approximately 76 per cent in data elements, and 69 per cent in disc storage. This is substantially less than the reduction initially hoped for. At the outset of the project, the goal was to reduce the amount of data for the 1 : 250 000 scale data set to at least one-sixteenth to one-twentieth of the amount required by

Table 3.1 Part of the 1 : 250 000 feature selection table (from Brown 1987a)

Description	EMR code	Data type	For 1 : 250 000 file:			
			Drop	Keep	Recode	Retype
Aerial cableway	2530	Line		x		
Aerodist tie point	1807	Line	x			
Airfield (250 000)	3085	Point		x		
Airfield (50 000)	3040	Line			3085	Point
Airfield position approx.	3084	Point		x		
Airport	3683	Point	x			
Airport runways	3020	Line			3683	Point
Alkali flat	5090	Line	x			
Amusement park	7680	Line	x			
Anchorage for large vessel	3090	Point	x			
Anchorage for small vessel	3100	Point	x			
Anti akoustikos	7120	Line	x			
Approach point	1802	Line	x			
Aqueduct	8100	Line		x		
Aqueduct underground	8110	Line		x		
Arena	4330	Line	x			
Arrow	9990	Line	x			
Arrowhead	5560	Point	x			
Astronomic monument	1415	Point	x			
Ballpark	7650	Line	x			
Barn	4050	Point	x			

Table 3.2 Summary of data reduction amounts for the 1 : 250 000 pilot project (from Brown 1987b)

File	Map sheet number							
	76O/9		76P/10		76O/15		76O/16	
	No. of elements	K bytes	No. of elements	K bytes	No. of elements	K bytes	No. of elements	K bytes
Original	14 577	3 442	13 067	3 571	81 174	8 387	61 802	6 346
After extraction	12 115	2 794	10 315	2 787	59 455	6 097	44 606	4 539
After extraction and polyline simplification	5 832	1 485	5 157	1 424	6 444	1 654	4 237	1 256
% of original	40	43	39	40	8	20	7	20

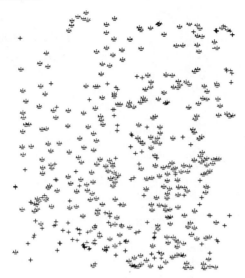

Fig. 3.2 Illustration of overly dense point symbol locations. This plot contains features selected for 1 : 250 000 and plotted at 1 : 250 000 from 1 : 50 000 map sheet 76O/9 (Brown 1987b)

the original 1 : 50 000 scale data sets. This reduction goal came from the fact that each 1 : 250 000 sheet contains the data from 16 1 : 50 000 sheets, and also from the fact that these data must be represented in one-twenty-fifth of the area of the original 1 : 50 000 map sheets.

Other difficulties with this purely algorithmic approach to generalization became apparent. Even with feature selection by table look-up, some features are still too dense when portrayed at the target map scale of 1 : 250 000. Figure 3.2 illustrates the problem. Another example of the difficulty with this algorthmic approach is depicted in Fig. 3.3. Even though the data in Fig.3.3(b) have been simplified using polyline simplification, they are still too dense to be useful at 1 : 250 000. Small streams and ponds need to be removed, further smoothing is required, and feature exaggeration is required in some places. The problem of interference between the hydrography and hypsography when combined on the final target map would have presented further problems.

The approach taken for this 1 : 250 000 pilot project is not the only one possible. Several different frameworks for generalization have previously been established (e.g. Beard 1987; Brassel 1985; Rhind 1973) and these are reviewed in Chapter 2 of this volume (McMaster 1991). Elimination of small lakes, ponds, and streams from the 1 : 50 000 hydrography could probably have been done using an automated elimination algorithm or the coarsen procedure used by Beard (1987) and Chrisman (1983). Interference detection between hydrography and hypsography, with its subsequent displacement, could possibly have been handled using Christ's (1978) or

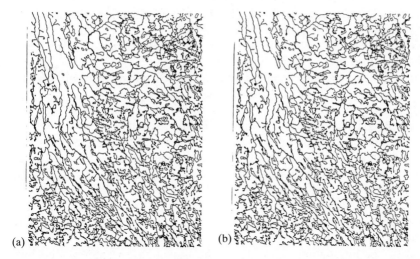

Fig. 3.3 (a) Not-to-scale plot of original hydrography of 1 : 50 000 data set for map
sheet 76O/9; (b) plot of hydrography for 76O/9 at same scale as (a) after
polyline simplification (from Brown 1987b)

Nickerson's (1988a) algorithms. No known algorithm would, however,
handle the required feature exaggeration.

In summary, the 1 : 250 000 pilot project illustrates the types of
generalization activity which a purely algorithmic approach can perform. It
also points out the difficulty of performing automated cartographic
generalization algorithmically for large, heterogeneous map data sets.
Evaluation of the success of the results of an algorithmic approach
invariably lies with an expert cartographer's judgement. In this project, the
manually generalized 1 : 250 000 maps already existed, and the fact that the
automatically generalized data did not compare well with the existing
1 : 250 000 product was immediately apparent. One promise of the rule-
based approach is that it can capture the essence of what good cartographic
generalization is and how it is performed. A much more effective automated
cartographic generalization system will hopefully emerge when this rule base
is combined with the known cartographic generalization algorithms.

Manual generalization at CCM

Approximately 250 man-hours were required to complete the manual
generalization of one 1 : 250 000 scale map sheet (no. 86A) from 16
1 : 50 000 scale map sheets. Map sheet 86A is in a largely uninhabited area
of Canada's Northwest Territories, and generalization of hydrography and
hypsography were the two major tasks. The major steps involved in the
generalization are as follows:

1. Assemble the original 16 1 : 50 000 sheets of line-work for both hydrography and hypsography.
2. Photo-reduce all 16 sheets to a positive image (black lines on clear background) at a scale of 1 : 125 000.
3. Produce two mosaics (one for hydrography, one for hypsography) by cutting and taping the 16 reduced sheets together.
4. Blue-line a sheet (1.22 × 1.07 m) of scribe-coat material with the hydrography mosaic. This is called the blue base.
5. Scribe the hydrography directly on the blue base using scribe tips which are double the line weight of the scribing needed at 1 : 250 000 (plus 0.001 inch (0.0254 mm)). Feature elimination, smoothing, agglomeration, and exaggeration all take place here.
6. Red-line a sheet of scribe coat (called the brown base) with the hydrography mosaic and blue-line the same sheet with the hypsography mosaic.
7. Scribe the hypsography directly on the brown base using double-width scribe tips. A minimum line separation of 0.005 inch (0.127 mm) at the published scale of 1 : 250 000 is adhered to. This scribing is done on a light table with the already scribed blue base placed underneath so that line interference can be detected and displacement of the hypsography performed to maintain the minimum line spacing.
8. Scribe a marsh key on a scribe coat blue-lined with the hydrography mosaic. This will normally require agglomeration, simplification, and/or elimination of marshes on the hydrography mosaic.
9. Prepare the black base containing spot elevations, marks for eskers and other symbology that was not already included on the blue or brown bases or on the marsh key.
10. The map surround, topographic names, and grid lines are added to make a final package for printing after the map sheet is inspected, reduced (to the 1 : 125 000 final scale) and has reproduction negatives made.

Figures 3.4 and 3.5 show a small portion of map sheet 86A/4 and 86A for which this process was completed.

The above process (hereafter called the manual process) is successful for two main reasons. First, the generalization is done at a scale which is twice the target scale of 1 : 250 000 using the original 1 : 50 000 line-work as a guide. All the necessary details are visually available for making the necessary generalization decisions. Second, the generalization is done directly in one step using the scribe tool (at approximately double width). The first reason is especially important; the intermediate scale of 1 : 125 000 is where most generalization decisions are made. It is considerably easier for the cartographer to visualize what the final 1 : 250 000 result will look like at this intermediate scale rather than at the original 1 : 50 000 scale.

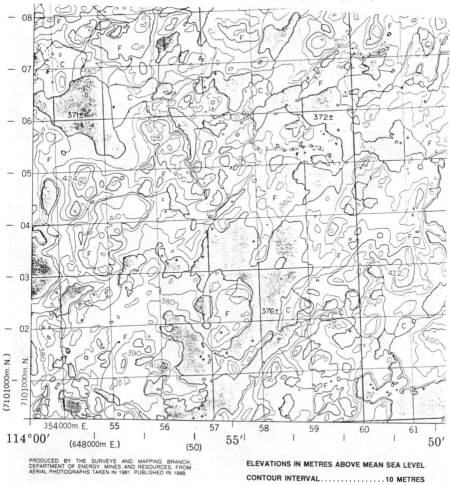

Fig. 3.4(a) Example south-west corner of 1 : 50 000 scale map sheet 86A/4

Rules for generalization

An overall guiding rule used when performing the above manual generalization is the minimum size rule. This can be stated as follows:

Rule R1: If a feature would be less than the designated minimum size for this feature at the target map scale, then it can be eliminated.

For the 1 : 250 000 generalized maps at CCM, this rule is embodied in the so-called 'derivation guide'. A reduced copy of this guide is shown in Fig. 3.6. This guide shows graphically the minimum size for various features at 1 : 250 000, 1 : 125 000, and 1 : 50 000. The guide is also printed on clear plastic so it can be placed directly over the 1 : 125 000 mosaic, and will

Fig. 3.4(b) South-west corner of generalized 1 : 250 000 scale map sheet 86A shown at approximately the same scale as (a). Note that this version is from a colour proof; it has not been inspected for errors

clearly show if features meet the minimum size requirement (EMR 1989). These minimum sizes are summarized in Table 3.3. These minimum sizes are not hard and fast, but serve as a general rule of thumb. For example, if a lake is not circular (as implied by Table 3.3), a judgement call is made by the cartographer as to whether the lake is less than the required size.

One exception to rule R1 is

Rule R1–E1: If there is not much drainage in the surrounding area, keep a lake feature which is smaller than the minimum size.

Another rule regarding emphasis of features is

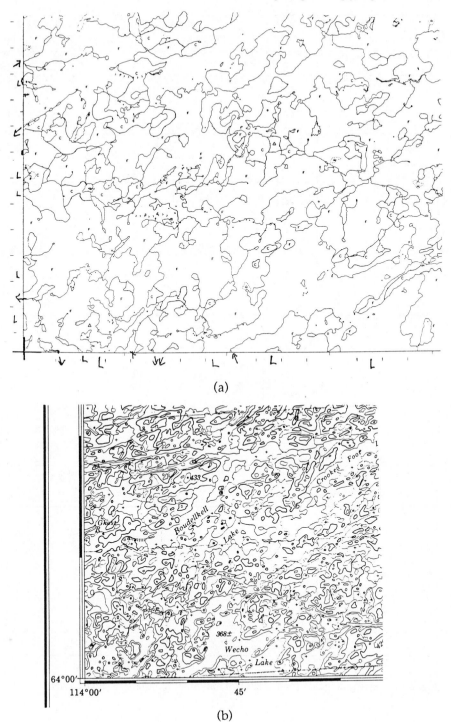

(a)

(b)

Fig. 3.5 (a) South-west portion of 1 : 125 000 hydrography mosaic for 84A; (b) south-west portion of 1 : 250 000 scale generalized map sheet 86A

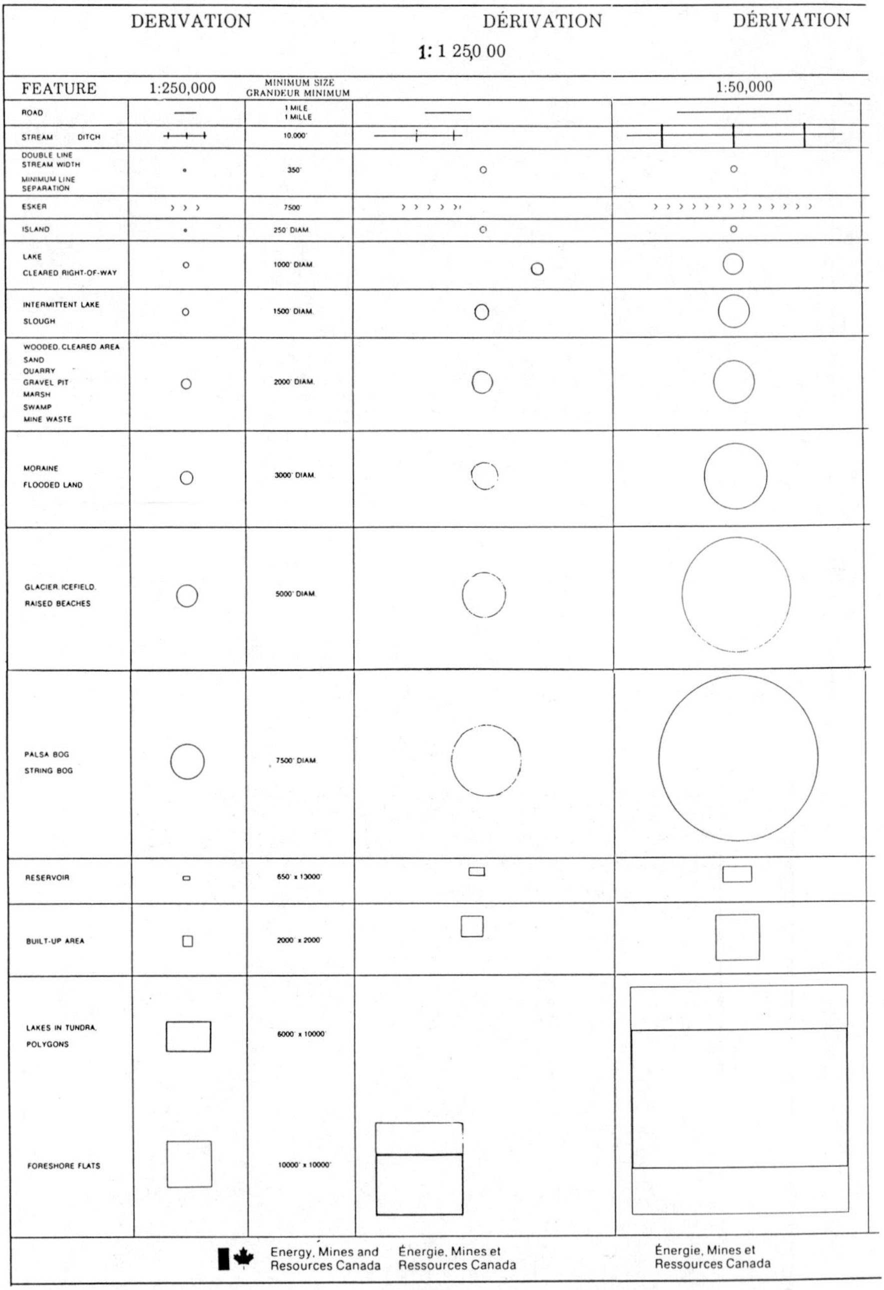

FEATURE	1:250,000	MINIMUM SIZE GRANDEUR MINIMUM	DERIVATION 1: 1 250 00	1:50,000
DERIVATION			DÉRIVATION	DÉRIVATION

Fig. 3.6 Derivation guide for 1 : 250 000 minimum feature size (EMR 1989)

50

Table 3.3 Minimum feature sizes for inclusion on a 1 : 250 000 scale map

Feature	Minimum size
Road	One mile (1.6 km)
Stream or ditch	10 000 feet (3048 m)
Double line stream width: minimum line separation	350 feet (106.7 m)
Esker	7500 feet (2286 m)
Island	250 feet diameter (76.2 m)
Lake or cleared right-of-way	1000 feet diameter (305 m)
Intermittent lake or slough	1500 feet diameter (457 m)
Wooded or cleared area, sand, quarry, gravel pit, marsh, swamp, mine waste	2000 feet diameter (610 m)
Moraine, flooded land	3000 feet diameter (914 m)
Glacier, icefield, raised beaches	5000 feet diameter (1524 m)
Palsa bog, string bog	7500 feet diameter (2286 m)
Reservoir	650 by 1300 feet (198 by 396 m)
Built-up area	2000 by 2000 feet (610 m)
Lakes in tundra, polygons	6000 by 10 000 feet (1829 by 3048 m)
Foreshore flats	10 000 by 10 000 feet (3048 m)

Rule R2: Exaggerate (by up to 1/3) prominent features which would be visible from the air.

As mentioned above, the rule forcing displacement to occur is

Rule R3: Maintain a minimum line separation at the published scale.

This minimum line separation is 0.005–0.01 inch (0.127–0.254 mm). This rule causes problems in urban areas where rivers, roads, railroads, transmission lines and contour lines may all lie nearly on top of one another in some places. In certain congested areas, the widths of the lines representing some features may be reduced by 0.001 inch (0.275 mm) to accommodate rule R3.

In general, hydrography is placed first with hypsography and other features being displaced to accommodate its location. This is another general rule stated as

Rule R4: Place hydrography first, with other features displaced (if necessary) to accommodate the hydrography location.

Rule R4 is a natural statement of the fact that hydrographic features play an important role in the interpretation of a map. Where many features overlap (thus violating rule R3), it may be necessary to ignore rule R4. In these situations, the following rule is used:

Rule R4–E1: Place straight-line features (e.g. railroads and transmission lines) first, with roads and edges of lakes or rivers being displaced with respect to them.

This rule (R4–E1) represents a commitment to keeping feature representations true to their physical shape on the ground. It was discovered that some cartographers preferred to stick with rule R4 while others used R4–E1 in highly congested areas.

Two other rules related to double-line features are

Rule R5: Dual highways with a median strip are shown as a single dual-line feature unless the median becomes wider than 500 feet (52.4 m); in this case, depict the highway as two single-line features following the highway location.

Rule R6: Avoid the 'sausage' effect when generalizing rivers.

Rule R6 means that a generalized river should be shown as having two distinct riverbanks, or as a single line feature, and not as a combination of both.

Besides these rules (gleaned by interviewing cartographers), many rules about construction of 1 : 250 000 maps are contained in the *Manual of Compilation Specifications and Instructions* (EMR 1974). For example, the following rules are concerned with built-up areas:

Rule R7: Villages or settlements up to 500 population will be shown with an open circle 1000 feet (304.8 m) in diameter. Towns of 500 to 2000 population will be shown with an open square 1500 by 1500 feet (457.2 m) in size. Towns of over 2000 population, if symbolized, will have the square shaded in.

This is vastly different from the 1 : 50 000 maps where most individual buildings are shown. Larger built-up areas are shown with a red urban-tint, and follow the rule

Rule R8: Only railways, main arterial roads and pertinent hydrographic features are shown within the urban-tint area at 1 : 250 000.

A significant difficulty is to determine which of the roads are the main arterial roads. Normally, these are chosen as the straightest roads within the urban area, with consideration given to their connection with other highways, bridges, and tunnels. Outside of urban areas, main arterial roads are those which are numbered on provincial road maps.

In summary, there are a myriad of rules used for manual generalization. Almost all of them are used to produce a consistent, understandable, readable final version of a 1 : 250 000 map. Dense map information normally corresponds to dense population centres, and this is where significant 'cartographic licence' sometimes comes into play. It is interesting to note that manual cartographic generalization of 1 : 250 000 map sheets was done under contract in 1989 for a cost of approximately $6500 (Canadian dollars) per sheet.

Automation

Can these rules and algorithms be combined into an automated cartographic generalization system? To a large extent, this depends on the underlying structure of the digital data representing the source (1 : 50 000) maps. Digital map data are commonly stored as different layers (e.g. hydrography, height, transportation, names) which are not linked in any way to one another. As seen above, successful manual cartographic generalization relies on visualizing all of the data in an area so that generalization decisions can be made. Figure 3.7 shows a possible architecture for an automated cartographic generalization system. There are several important features of this architecture, including completeness of the source data and assumptions about the operation of computer-assisted generalization.

The assumption of completeness of the source data files means that files have a complete topological structuring among all layers of data, and that the cartographic features therein are represented as complete objects. Complete objects are those containing all of the details necessary to make accurate decisions regarding their generalization. For example, generalizing the built-up areas of a city into possible single-tinted polygons representative of built-up areas at a smaller scale (see the example in McMaster 1991) requires that the data representing the objects in the built-up areas (e.g. houses, streets, other buildings) are somehow linked to the city object. Similarly, generalizing a multi-lane divided highway with its interchanges into a single line representation with simple squares representing the interchanges requires that the interchanges form part of the complete road object. An object-oriented data model (as proposed in Nickerson 1988b and expounded in Guptill 1990) is one possible data organization which would be useful for representing complete objects to assist in automated decision-making for generalization.

The second assumption concerns the generalization process itself. Generalization is assumed to be computer-assisted through the use of a 'generalization workstation', rather than completely automatic. This is essential as algorithms which effectively replicate the powerful human visual processing capabilities are still not possible. The rule-driven generalization algorithms operate on the complete merged results of all the source files providing data to the target map. This corresponds to the manual generalization practice of performing the generalization on a complete mosaic of all 16 source sheets, as explained above.

The so-called 'rule-driven algorithms' correspond to rules similar to those mentioned in the previous section, which can, if required, invoke procedural algorithms. For example, one rule might be

IF   an area of feature overlap is detected   [3.1]
THEN attempt to displace the overlapping features to meet the
          minimum spacing requirement

The left-hand side of this rule might invoke a raster algorithm for feature

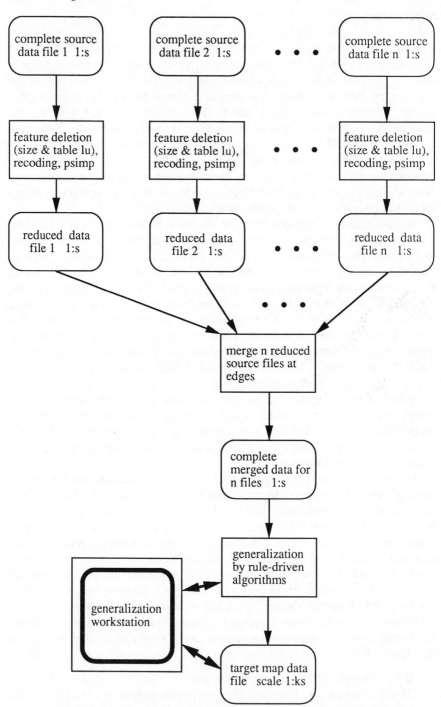

Fig. 3.7 An architecture for computer-assisted cartographic generalization. The term 'psimp' indicates polyline simplification. The term 'table lu' indicates table look-up. The terms '1:s' and '1:ks' indicate generic scale designations.

interference detection, while the right-hand side might invoke a vector displacement algorithm to remedy the situation. The minimum spacing requirement is not necessarily a fixed value as discussed in rule R3 above. It is important that these high-level rules be encoded in a grammer similar to that used by modern expert system shells. This allows for much easier understanding and modification of the rules by both system developers and users.

Summary

The cartographic generalization process relies heavily on human visual processing capabilities. A combined rule-based and algorithmic approach to automated cartographic generalization seems to hold promise for success in the future, given that the rule base can be established. If this process is to be even partially automated, map data must be structured so that complete objects (such as a road with its interchanges) are available for the rules to reason about. Some rules and techniques for cartographic generalization which have been developed at the CCM have been identified. The knowledge engineering task for capturing a complete set of cartographic generalization rules is not yet complete, but an effective start has been made.

Acknowledgements

The assistance of the Natural Sciences and Engineering Research Council (NSERC) of Canada in support of this research is gratefully acknowledged. I would also like to thank the management and staff of the CCM in Ottawa for amiably accommodating my many questions. The anonymous referees are acknowledged for their comments which led to a better chapter.

Part II

Data modelling issues

4

Representing geographical meaning

Timothy L. Nyerges

Introduction

Model building in geography as well as other social and natural sciences has always suffered from a lack of applicability, many models often being called unrealistic simplifications of reality. Brassel and Weibel (1988) as well as this author contend that digital map generalization is part of the broader field of spatial model building, the focus here being geographical model building. The concern in this chapter deals with techniques to support development of practically adequate database models of (some portion of) reality for digital map generalization. A database model is a design (framework) for interpreting a specific database, but is not the database itself. The concepts presented here extend to descriptive, explanatory, and predictive geographical modelling in general (Nyerges 1991b).

It is argued here that practically adequate models of (some portion of) a geographical reality, i.e. models useful in carrying out decisions about that reality, must include knowledge of geographical meaning. Understanding why we need to consider geographical meaning when solving geographical problems is an age-old topic in geographical investigation. In lieu of a lengthy discussion concerning the philosophy of science, the author contends that more meaningful database models are more realistic database models (Nyerges 1991b).

One of the reasons why experienced cartographers who perform map generalization have not (and perhaps cannot) fully describe how generalization is done is because they have never had a means systematically to document the knowledge they use to perform a generalization task. The fundamentals of that knowledge concern meaning. Meaning involves context as well as a definitional ontology (Nyerges 1991b) of geographical structure and process. Using a high-level, somewhat formalized language that can be converted to a computational form, is probably the best way to proceed with examining the nature of the map generalization problem. This chapter represents another step towards that goal.

Maps are considered representations of (some portion of) a geo(graphical) reality. In a digital context such a representation is a geographical database. When digital symbols that are visually oriented (or other sense-oriented) are added to a geographical database we call it a cartographic database. Consequently, a geographical database is the core of a cartographic database. Since a geographical database is the core, geographical meaning has an impact on cartographic meaning. Geographical meaning is drawn from meaning of the referents in the world, rather than from symbols that appear on a map (Board 1984). It is the task of a cartographer as a database designer to provide enough meaning to convey (a nature of some) reality. In providing more meaning, a database designer most likely would be enhancing the database since more meaning reflects more of (some) reality.

Geographical databases as digital representations of human conceptions of the world currently rely on aspects of space, theme, and time to convey meaning. Geographical databases consist of more than just points, lines and areas with attendant attributes. They consist of implied geographical meaning given in terms of high-level concepts such as definitions and the relationships among phenomena that describe geographical structure and process. We need to extend our treatment of geographical databases to include knowledge of geographical meaning if we are going to progress with digital map generalization. Data management capabilities must be enhanced with knowledge representation capabilities. This can be accomplished through enhanced data management architectures that make use of geographical information abstractions.

Abstraction and generalization are a fundamental part of cartographic activity, both for map creation as well as map use (Brassel and Weibel 1988). Dent (1990) views abstraction as a process of isolating the important aspects of a topic from a wide range of concerns in the world, and including these important aspects in a map message. In a digital environment this means isolating important aspects of the world and representing them in an appropriate database design, and subsequently including them in a geographical database. Explicit knowledge of abstraction is critical for map generalization because such knowledge provides the context for why certain phenomena are represented in a geographical database in the way they are as opposed to other representations. A representation of meaning for and in a geographical database could amplify the intelligence of data in support of digital map generalization (Weibel 1991, Ch. 10 this volume).

Digital map generalization, although potentially composed of many kinds of operations, is a process of reducing information volume while at the same time preserving the significant information to be portrayed. To date, digital map generalization has not progressed much beyond surface filtering, line simplification, and point pattern reduction. Why has progress in digital map generalization been so slow? Much of the difficulty with digital map generalization is that knowledge of the abstraction process in developing a geographical database has been lost, is not available, or at least not retrievable in current systems. Knowledge of geographical meaning as part of the abstraction process is seldom captured since database development of primitive points, lines, and areas with attendant attributes has been a tremendous job in itself. In addition, software systems at the current time do not have the ability to capture such information. Much of that problem is due to the fact that concepts and techniques have not been formalized to the extent where such software tools are feasible. This chapter addresses such issues.

To address the lack of knowledge, one immediate concern is with what knowledge do we capture? Another concern is how to represent the knowledge if we could capture it. This chapter addresses those concerns with special focus on knowledge about the meaning of geographical phenomena. Meaning in this context consists of an internal meaning and an external meaning. An internal meaning deals with a scientific, dictionary-based meaning, composed commonly of necessary and sufficient conditions that include spatial, thematic, and temporal aspects of information. An external meaning is one that includes significant relationships among phenomena. The relationships are derived on the basis of various spatial, thematic, and temporal aspects of entities in the database. Those relationships and entities in the database represent the corresponding relationships and phenomena in the world.

Geographical information abstractions are constructs that use both definitions and geographical relationships in order to develop the conceptual content of databases. Rhind (1988) states that we have not done very much with the conceptual content of geographical information in a geographical information system (GIS) context, as we continue to focus on primitive representations of geographical phenomena in terms of simple points, lines, and areas with their attendant attributes. This is not to say that point, line, and area primitives characterized using attributes are not important, quite the contrary. The conceptual content is stored as metadata, that is the data which are used to describe the meaning of the primitive data elements. By focusing on conceptual content, as higher-order data composed of definitions and relationships, it is possible to develop databases that are 'intelligent' (Parsaye *et al.* 1989) in order to make further progress with digital map generalization.

One of the major problems with the conceptual content of geographical databases, and thus meaning in databases, concerns how to represent this kind of information. The problem requires suitable knowledge representation techniques which are practically adequate for the task at hand to represent geographical meaning. When an information abstraction is practically adequate for map generalization it characterizes the nature of reality such that expected outcomes of generalization decisions are sufficient (Sayer 1984), i.e. outcomes are practical and useful solutions to particular map generalization problems.

This chapter proceeds by first clarifying the differences in geographical information abstraction and map generalization for the benefit of mutually supportive research on both topics. That discussion is followed by an examination of four types of geographical information abstractions for cartographic generalization using an example of geographical generalization presented by Mark (1989). Knowledge representation techniques to represent information abstractions are then discussed in the context of the example. A brief discussion of implementation considerations follows the section on knowledge representation techniques. The last section presents concluding comments and directions for continued research.

Differentiating information abstraction and map generalization

Geographical information abstraction is similar to and at the same time different from map generalization. Both information abstraction and map generalization focus on reduced complexity in information, but in different ways. An information abstraction process retains information in a database by temporarily suppressing it from use; whereas a map generalization process has traditionally eliminated information from a database because a new version is made. With geographical information abstraction the information is accessible upon request, but with map generalization the information is inaccessible because it has been eliminated from the database.

Geographical information abstraction mainly concerns managing geographical meaning in databases, and map generalization mainly concerns structuring map presentations. For these reasons, it is convenient and useful to separate geographical information abstraction and map generalization. This allows us to put research effort where the effort will do maximum good since confusion in concepts is counter-productive. However, it is useful and important to know where the two concepts influence each other for synergistic research purposes.

Geographical information abstractions are a way to represent the meaning of information in geographical databases. Such abstractions comprise the multiple conceptual hierarchies in a geographical database. These conceptual hierarchies give meaning to data objects specified in terms of spatial, thematic, and temporal data domains (Brassel and Weibel 1988; Sinton 1978). When the multiple conceptual hierarchies are taken together, they form a heterarchy of concepts. A heterarchy is the multidimensional knowledge framework of the concept meanings (more elaborate than a flat or even hierarchical semantic network). Those concepts are stored in the form of metadata representations, i.e. the definitions that provide meaning for the data as well as the relationships between data. The metadata representations are intended for heterarchical processing, i.e. from top to bottom, bottom to top and side to side for concept reference when required by generalization operations.

Geographical information abstractions stored, or derived as part of the information in a geographical database facilitate the use of geographical knowledge when creating a cartographic database. Mark (1989) encourages researchers to look at geographical generalization rather than being limited to cartographic generalization. Geographical generalization takes into consideration the influence of geographical phenomena on each other rather than starting with cartographical data as symbolized geographical data. Mark's concern for geographical generalization stems from his long-standing interest in the nature of geographical phenomena and their impact on cartographic, phenomenon-based data structuring (Mark 1979). That concern continues to be the focus of geographical entity structuring within cartographical databases (Nyerges 1980) in this research.

Brassel and Weibel (1988) use the terms digital landscape model (DLM) and digital cartographic model (DCM) to describe a geographical database and a cartographic database, respectively. A geographical database contains representations of geographical phenomena defined in terms of entities, whereas a cartographic database is a symbolized geographical database described in terms of graphical symbols.

Brassel and Weibel (1988) also distinguish statistical generalization for DLMs as being different from cartographic generalization for DCMs, and say that cartographic generalization is the more significant of the two for map generalization. Statistical generalization is described as an analytical process having to do with information content reduction in a database under statistical control, with both the source and target databases being characterized using an accuracy measurement. They describe cartographic generalization as having to do with a scale reduction in a map while preserving the representativeness of a map in a holistic, non-statistical fashion. Conceptual differences in a geographical database resulting from statistical generalization are discounted in their discussion, since the process is under statistical control. Thus, they assume that the conceptual differences between the original geographical database (DLM) and the resulting geographical database (DLM) are non-existent or minimized, having little influence on a cartographic database.

Although statistical techniques can be used to perform many operations, they are not yet sufficient to control loss of 'meaning' as contained in geographical information abstraction constructs. In fact, geographical information abstractions can assist with performing statistical generalization by helping to preserve the meaning of the resultant spatial distributions. This would involve representing concept lineage for spatial distributions. In doing so, selected characteristics are statistically processed, but whether the conceptual meaning of the spatial distribution changes depends to a significant degree on just what those characteristics are.

Geographical information abstractions provide meaning for the map generalization process. The conceptual process is depicted in a simplistic fashion in Fig. 4.1 whereby information abstraction precedes map generalization. Information abstractions carry the geographical meaning in DLM 'X',

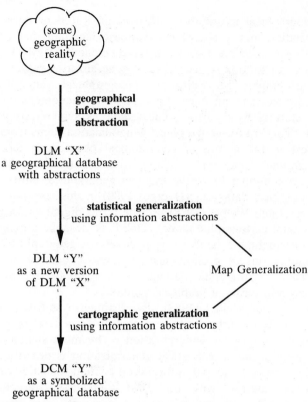

Fig. 4.1 Geographical information abstraction and its relation to map
generalization for a digital landscape model (DLM) and a digital
cartographic model (DCM)

hence DLM 'Y', and support the statistical and/or cartographic generalization
process. The focus in geographical information abstraction is on representation of the spatial, thematic, and temporal meaning of a particular database,
and is not concerned with transforming one representation into another, as
is the concern in the cartographic and statistical generalization processes.
Although statistical generalization looks similar to geographical information
abstraction as a process they are quite different in outcome. Geographical
information abstractions are stored in databases to allow temporary
suppression of detail, for example to enhance the semantics of DLM 'X' as
well as to retain the information lost in the transformation from DLM 'X' to
DLM 'Y' in Fig. 4.1. In this way, knowledge of spatial context at multiple
levels of resolution is retained and used in DLM 'X' for temporarily
suppressing the detail not needed in DLM 'Y' or DCM 'Y'. Using
knowledge of geographical meaning to suppress cartographic details is
fundamental to map generalization.

Using information abstractions in map generalization

To characterize the nature of the map generalization process Brassel and Weibel (1988) present a conceptual framework that includes the following steps: (1) structure recognition of objects in a source geographical database; (2) process recognition of how the structures are to be converted to structures in a target cartographic database; (3) process modelling involving a set of rules for sequencing the procedures that effect the change in structures; (4) execution of the rules; and (5) display of the cartographic image.

Structure recognition in map generalization involves identification of objects and aggregates of objects, spatial relations between objects, and establishment of measures of relative importance for these (Brassel and Weibel 1988). Brassel and Weibel (1988) state that structure recognition for cartography is rather poorly defined at present. Geographical information abstraction constructs are the repository for the conceptual information identified as part of structure recongition. Such abstractions represent, in part, the structural knowledge of geographical neighbourhood relationships important for structure recognition. In this context, a neighbourhood is taken to mean a set of entities and their relationships that act as a geographical unit for map generalization, i.e. a complex object identified for processing by a set of cartographic generalization operations.

The elements of the complex object influence each other in a given area sufficiently so as to be considered as a whole during the map generalization process for that area. Integration of the neighbourhood elements (based on space, theme, and time) takes on the form of a heterarchical abstraction network of concept types and objects. The abstraction heterarchy represents the meaning encoded by analysts who develop a database, and the meaning that is derived using neighbourhood-focused (spatial, thematic, and temporal) software operations.

Based on work with visual abstraction in visual understanding (Marr 1982), with visual routines in visual cognition (Ullman 1985), and with cognitive image schemas in cognitive linguistics (Lakoff 1987), it is asserted here that structure recognition proceeds as a cognitive process (whether implemented by machine or human) that employs cues at multiple levels of meaning. Part of the problem is that we must first know how to characterize a structure in some meaningful way before we recognize it. Meanings are not generated spontaneously, but evolve from what is already known. Structure recognition, hence structure identification, starts with global, significant characteristics, and then proceeds to more detail. The internal meanings given by definitions, and the external meanings given by spatial relationships, both employ all three aspects of space, theme, and time at the global level first, and then proceed to details. Use of the concept heterarchy provides the basis for geographical structure identification at global and detailed levels, and consequently cartographic structure identification.

Structure identification in cartographic (Buttenfield 1987) and geographical

(Mark 1989) data is an important topic in current research for the map generalization knowledge domain. Buttenfield (1987) explores the geometric structure of scale-dependent geometric structure of cartographic lines. She asserts that different features exhibit a different structure signature (see also Ch. 9 in this volume, Buttenfield 1991). This assertion suggests that feature meaning is rather important to an intepretation of geometry. Thus, characterizing the spatial structure component is a first important step, but we need to extend this to the thematic and the temporal components as well to understand more about the geographical structure to be generalized. Adding a thematic component to spatial structure involves adding the 'what the phenomenon is' component, approaching a truer description of geographical structure. This is suggested by Mark (1989) in his investigation of geographical generalization, commenting that the 'nature of the phenomena' has a lot to do with how the individual phenomenon, hence symbols, will be generalized in relation to the phenomena surrounding it.

Types of geographical information abstraction

Four geographical information abstractions are important in providing sufficient knowledge of meaning to perform structure identification. The four types of abstraction are **classification**, **association**, **generalization**, and **aggregation** (Brodie 1984; Schrefl, Tjoa, and Wagner 1984) as depicted in Table 4.1.

The four abstraction types are the same structural relationships identified by the author (Nyerges 1980) to support meaning in geographical information, and have been used by Schrefl, Tjoa and Wagner (1984) to compare and contrast expressiveness among several semantic data models. Semantic data models are frameworks for designing database management systems which incorporate more meaning than the traditional hierarchical, network, and relational models (Peckham and Maryanski 1988). Very little has been done with the abstraction types with regards to geographical information because knowledge representation techniques and efficient,

Table 4.1 Abstraction types for semantic data modelling (adapted from Schrefl, Tjoa, and Wagner 1984: 122)

Abstraction	Relationship	Relationship between
Classification	Instance-of ($\rightarrow$)	Entity $\rightarrow$ entity-type Entity $\rightarrow$ entity-class
Association	Member-of ($\rightarrow\rightarrow$)	Entity $\rightarrow\rightarrow$ entity-group
Generalization	Kind-of (=>)	Entity => entity Entity-type => entity type
Aggregation	Part-of (.) for entity (..) for definition	Entity . entity Entity-type .. entity-type

object-oriented implementation mechanisms have not yet been introduced into spatial data management. Although some researchers have considered spatial data management systems to be designed after semantic data models (Brodie 1984), storing geometric and topological relationships does not quite comprehensively deal with the nature of geographical meaning. All four abstraction types, classification, generalization, aggregation, and association, add something different to the nature of geographical meaning.

Before starting into a discussion of each of the abstraction types, a few thoughts of clarification are necessary for the term generalization, since it is used in both map generalization and in database abstraction. Generalization as a scientific endeavour has had a somewhat narrower interpretation in the database abstraction literature than it has enjoyed in cartography. The term generalization in the database literature follows closely the use in the philosophy of science literature more so than it does in cartography. The philosophy of science meaning of generalization involves 'a concept having a more general interpretation than some other concept with a more specific interpretation'. In cartography, the term generalization commonly applies to selection, simplification, classification, induction, and symbolization (Robinson *et al.* 1984), and should really be prefaced with the term 'map' or 'cartographic'. Therein lies the potential confusion for the reader.

The two uses, one in the map generalization literature and the other in the database abstraction literature, are both important here because databases as models of (some portion of) a reality and maps as displays of (some portion of) a reality are important for this discussion. However, since abstraction precedes map generalization, and because the focus in this chapter is on database abstraction, the narrower definition for generalization is implied in this section when the term is used without a qualifier, i.e. a generalization is a less specific rendition of a concept. This usage may cause cartographers some confusion at first; but we must remember that interdisciplinary work often results in some initial confusion, and a context helps make term usage clearer. The author has tried for years to come up with a better term for generalization in both the map generalization and database literature; however, the terms are so entrenched in the respective fields of cartography and computer science that many more years are required to bring sufficient clarification. In the following discussion 'cartographic generalization' is used to refer to the cartographic context, and 'generalization' is used for the database abstraction context.

Classification

As the first of the four abstraction types to be discussed here, classification is the process for determining categories of information included in a database. Definitions provide an easily communicated meaning as a basis for categorization. A definition of a river is a useful way to represent what is meant by 'river' when a database is to be designed and used. A definition

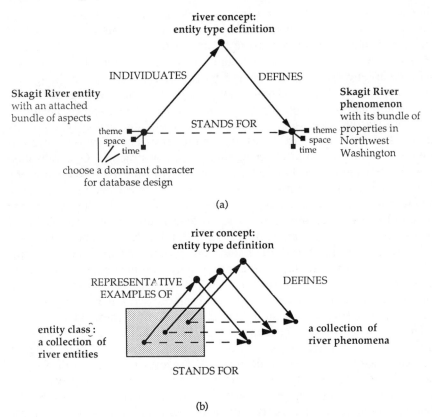

Fig. 4.2 (a) A meaning triangle for a river entity; (b) classification of river entities for database design

explicitly indicates how phenomena are to be interpreted, and is the basis for forming classes. Definitions are usually developed by specifying necessary and sufficient properties for referents in the world (Sowa 1984) as interpreted from some organization perspective such as the US Geological Survey (USGS). However, definitions based on prototypical examples and probabilities are also possible (Sowa 1984), these having a different epistemological basis for determining what constitutes a referent.

A meaning triangle (Sowa 1984; Nyerges 1991b) depicted in Fig. 4.2(a) provides a way of understanding the significance of geographical meaning as given by an entity type definition for a 'river' concept. A phenomenon, such as the Skagit River in north-west Washington State, is some portion of reality which we see, use for recreation, and otherwise experience. Such a phenomenon is described in terms of various properties as observed; primarily for geographical databases these include some attached, but inseparable, bundle of theme, space, and time. Fundamentally for geographical databases, a definitional perspective is a matter of deciding which one of a theme, space, or time aspect is dominant for an entity-type definition, and then adding to the definition using the other two aspects. None the less, all three aspects are required for database entities to maintain complete geographical phenomenon (feature) descriptions.

If the Skagit River is to be represented in a database then some aspects of theme, space, and time must be selected to individuate (Brachman 1979) the river concept, with the intended representation in the database called an entity. The same aspects of the river entity 'stand for' the properties of the phenomenon (as they exist in reality) but only indirectly through the definition, as is indicated by the dashed line. Any one of the theme, space, and time aspects is likely to dominate a focus on an entity in the database design process, as is shown in Fig. 4.2(a). The dominant aspect is usually reflected in a definition, and provides a focus for the intended meaning of a representation in a database.

For brevity, a definition for a particular entity class is called an entity type, also sometimes called a concept type (Sowa 1984). Signifying something as a 'type' implies that a reason exists for constructing the type, and that reason is central to the meaning of the type, given in terms of a definition. Consequently, an entity class is a collection of representations that stands for the collection of phenomena in the world (see Fig. 4.2(b)), explicitly given a meaning in terms of an entity-type definition. A rectangle is commonly used to depict an entity class in a database model, as is the case in Fig. 4.2(b).

Classification is not new to GIS and has been the fundamental abstraction type used for categorizing data, but with little formal treatment since it is fundamental to human experience. However, two perspectives currently exist on the process of creating the classifications. Burrough (1986: 13) suggests that a 'map is a set of points, lines and areas that are defined both by their location in space with reference to a coordinate system and by their non-spatial attributes'; a perspective the USGS supported in their original development of the digital line graph (DLG) data. The DLG database model emphasizes a geometry-dominant perspective. That perspective differs from the more recent perspective for developing data as in the USGS DLG–enhanced (DLG–E) as described by Guptill and Fegeas (1988). The DLG–E database model emphasizes a theme-dominant approach that supports a feature-based model, a feature being an entity and its object representation in a database. In the feature-based model theme is fundamental, such that a feature which is defined phenomenologically, will supersede how the feature is to be represented geometrically as a point, line, area, or volume. Thus, the geographical definition of an entity class (for a world referent) based on its thematic character is perhaps the more stable construct for building a database.

Association

An example of a database model that employs various entity classifications and associations for a database is depicted in Fig. 4.3. An association relates entities by virtue of a member-of-set relationship (Brodie 1984). Two or more entity classes together with the corresponding relationship(s) are called an association. Associations are significant for composing geographical neighbourhoods, i.e. a collection of entities and relationships that have an influence on each other. The relationships as depicted follow from influences presented by Mark (1989) in his example of geographical generalization, but are embellished with others as well. Any one of the influence relationships is useful for describing a neighbourhood. However, such neighbourhoods are determined by problem-specific conditions. Consequently, each of the geographical influences is interpreted in terms of a set of conditions for spatial, thematic attribute, and/or temporal aspects as criteria for defining a neighbourhood.

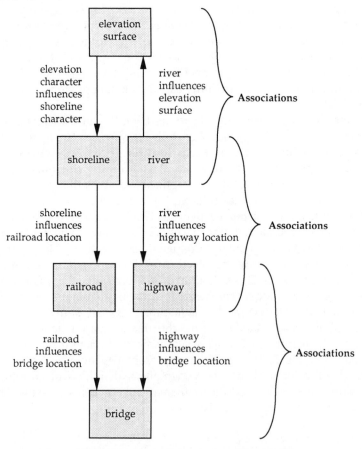

Fig. 4.3 A database model depicting classifications and associations for geographical neighbourhood influence

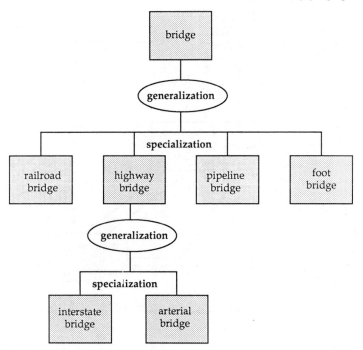

Fig. 4.4 Generalization/specialization for bridge entity class

In Fig. 4.3 problems exist with the interpretation of almost all the entity classes. It is possible to expand each of the entity classes into a database model on its own, but space does not permit such development here. However, two are of particular interest. The 'bridge' and 'elevation surface' classes lack specificity to understand clearly the nature of the 'influence' as caused by different entities. Here we examine only briefly the 'bridge' and 'elevation surface' classes which point out the usefulness of two other abstraction types, generalization and aggregation.

Generalization

A generalization is a process of making entity classes less specific by suppressing characteristics that describe the class (Figure 4.4). In a reverse sense, a general concept for 'bridge' can be specialized into different types of bridges, e.g. highway, railroad, pipeline, and foot, by adding qualifying characteristics to the more general concept. Consequently, specialization is the inverse of generalization. A generalization as a database abstraction is expressed as a concept-type hierarchy that represents levels of concept specificity. Furthermore, a generalization as an abstraction type organizes levels of both entity instances and entity-type definitions.

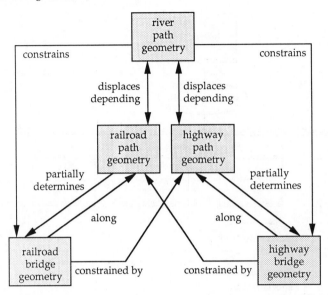

Fig. 4.5 Association for describing geographical neighbourhood influence

Database processing takes advantage of specializations because more details are available to direct the processing. Thus far we have been concerned with a thematic attribute-dominant specification (ADS) of a database model. An ADS is more concerned with meaning as given by thematic attributes rather than with its spatial meaning (structure). At this place in the discussion it is convenient to introduce a space-dominant specification (SDS) that provides locational focus. Entity geometry is specified in the SDS to detail further the character of entities as they are (to be) represented in a database.

Entity classes from Fig. 4.3 and 4.4 are specialized according to geometric representations, as depicted in Fig. 4.5. In addition, the spatial relationships of Fig. 4.3 are detailed further, providing qualified character for the spatial influences. Rivers impact or are impacted by railroad path and/or highway path geometry depending on the focus of the database design, hence displacement depends on setting appropriate conditions. In addition, the river path geometry constrains the railroad and highway bridge geometry since bridges cross over rivers. The railroad path geometry partially determines the railroad bridge geometry by forcing the railroad bridge to lie along the railroad path.

Aggregation

The 'elevation surface' concept in Fig. 4.3 also needed further semantic clarification. In Fig. 4.6 (p. 76) an aggregation abstraction collects localized surfaces together into a supra-order entity class called elevation surface, and at the same time identifies the inverse, i.e. a disaggregation of elevation surface to be localized terrain surface. This is necessary to differentiate more easily the role of a river and a stream during map generalization. A river meanders across an elevation surface as a major influence on the terrain, while the course of a stream is often determined by the character of the terrain surface. In addition to describing the individual entities, an aggregation abstraction must also manage the definitions of entity classes that are created as composite definitions.

The four types of abstraction introduce geographical meaning into databases, but only through a software framework that supports the abstractions. Nyerges (1987), Goodchild (1987), and Peuquet (1988a) make note of the deficiencies of spatial data models in representing higher-level abstractions other than points, lines, polygons, and images, particularly in vector-oriented models. They all suggest that further research into those deficiencies involves both data description constructs and operations in a data model. However, the next question to ask is whether incorporating such constructs on-line would be useful in a data model, or should such constructs simply be a guideline for database design? If such constructs are not included in a data model then knowledge processing would not have access to this information. This has been part of the problem for digital map generalization in the past. Too little on-line information about the meaning of geographical phenomena results in primitive processing, hence slow progress in digital map generalization. Database guidelines that are not fully automated are nice as reference documents, but for on-line digital map generalization knowledge representation techniques are required.

Knowledge representation techniques for information abstractions

The following discussion builds upon the fact that current spatial data models carry limited knowledge of meaning, and begs a somewhat epistemological question. What kind of knowledge representation techniques are useful for implementing the geographical information abstraction types regardless of whether the abstraction types are implemented as part of the database or as part of the metadata describing the database?

Sowa (1984: 304–5) lists seven knowledge representation techniques useful for implementing database semantics: type hierarchy, functional dependency, domain role, definition, schema, procedural attachment, and inference rule. As with any knowledge, it is argued here that some is structural knowledge, some is process knowledge, whereas still other knowledge comes as facts. Type hierarchies, functional dependencies, domain roles, definitions, and schemata compose the structural knowledge representation techniques. Procedural attachments and inference rules compose the process knowledge representation techniques. The entity class primitives of the database compose the asserted facts. The structural and process knowledge representation techniques are described below.

Structural knowledge representation techniques

Type hierarchy

Entity types and entity classes in a database are ordered according to levels of semantic generality. The ordering indicates that some entity classes have fewer details to describe them. A lattice structure implements a type hierarchy, with lower-order types inheriting the characteristics from higher-order types. A lattice is a multiconnected graph with special properties. Multiple inheritance in a lattice is possible, i.e. a lower-order type may share characteristics from two or more higher-order types.

Functional dependency

Dependencies indicate which data characteristics are primary or secondary referents. Primary data referents are keys or independent variables. Secondary referents are functionally dependent on primary keys. Quantizers indicate how many of one referent are related to another, thus whether a function is many-to-one, one-to-one, or many-to-many. Specific locational identifiers, and special thematic attribute values are possibilities for keys. Procedures are used to derive relationships once keys have been established.

Domain role

Roles apply to entity classes, and therefore entities, in specific context domains. A role is a way of interpreting the interaction of an entity in relation to another entity, for example a river in relation to other transportation entities and a river in relation to other parts of a drainage network. A domain is a collection of values that are valid for the relation between the entity classes. Roles establish functional relationships between one entity class and another by virtue of specific values having some well-defined interpretation. An entity may take on multiple roles depending on the nature of the relations with other entities, for example as in the case of the drainage and transportation contexts.

Definition

Three types of definitions include: (1) classical – entity inclusion by necessary and sufficient conditions (formally called genus supertype and characteristic differentiae); (2) prototype – entity inclusion by best example; and (3) probabilistic – entity inclusion by statistical commonalty of characteristics. The most common formal definition for entity types and relationships is given by necessary and sufficient conditions consisting of thematic, geometric, and temporal characteristics. It is possible to form aggregation types and entity types from the other entity types, resulting in compound definitions.

Schema
A schema (plural of schema is schemata) describes the normally occurring, or default, roles that an entity type plays with respect to another entity type. In some instances a schema provides background information in terms of defining characteristics other than primary type given by definition. A schema is a structure that contains slots. A slot holds a value, another schema reference, a rule for interpreting data as facts, or an attached procedure for deriving other data values (Armstrong 1991, Ch. 5 this volume).

Process knowledge representation techniques

Attached procedure
Attached procedures bound to a schema indicate how external procedure calls compute the referents of entity classes. A set of criteria for a target structure in cartographic generalization determines when a procedure call is invoked. Procedure expressions may describe computed functions, relationships stored in the database, or virtual relationships computed as needed. Attached procedures take the form of cartographic generalization operations as discussed by Shea and McMaster (1989) and McMaster (1991, Ch. 2 this volume).

Inference rule
An inference rule represents a stage in reasoning based on explicitly stored knowledge of entity classes and their instances stored as objects in a database. Inference rules derive relationships between entity objects through reasoning rather than algorithmic computation. Inference rules describe how the structural knowledge should be used in structure recognition (Buttenfield 1991, Ch. 9 this volume). They are also used to detect violations of the constraints for cartographic generalization operations.

Table 4.2 Knowledge representation techniques used to implement abstraction types

Knowledge representation technique	Abstraction type			
	Classification	Association	Generalization	Aggregation
Type hierarchy			•	
Functional dependency	•			
Domain role		•		•
Definition	•	•	•	•
Schema		•	•	•
Attached procedure		•	•	•
Inference rule	•	•	•	•

If the previously discussed seven knowledge representation techniques are fundamental for representing database semantics, then they should be useful for implementing the geographical information abstractions. Table 4.2 cross-lists the four abstraction types against the knowledge representation techniques important for implementing each abstraction. The 18 entries in Table 4.2 are a start for determining what knowledge techniques are useful for representing geographical database meaning using the abstraction types. Other researchers may make different choices, such choices being a matter of design considerations. The following discussion includes justification for the choices made here, taking each abstraction type in turn. Reference is made to the example presented in Fig. 4.3–4.6.

Classification

Functional dependency used for classification

Data characteristics within each entity-type definition are accessed according to primary and secondary reference. Dependencies state the primary and secondary priority of the attributes in each entity-type definition. These dependencies are important in structuring the meaning of the classes, i.e. the definitions which describe the class meaning. Each of the classes in Fig. 4.3–4.6 requires that the dependencies between primary and secondary attributes be specified. Space is the primary consideration in these descriptions although only Fig. 4.5 makes this explicit. Thematic attributes, except for those in the definition of an entity type, are secondary. In

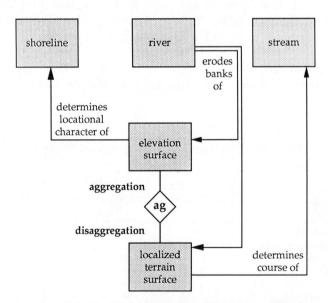

Fig. 4.6 Aggregation and association abstractions. Elevation surface is an aggregation abstraction of localized terrain surface

addition, access to individual objects as well as neighbourhoods of objects is implemented through the primary and secondary references. The quantizers for the influence of one entity object on another are specified as one-to-one, many-to-one, or many-to-many.

Definition used for classification
An entity type is a definition for an entity class, i.e. how one distinguishes the meaning of one class from another. Thus, a definition is specified in terms of characteristics used to recognize one entity as being similar to or different from another entity. The characteristics are often primary attributes for describing the referents to be included in the class. Each of the attributes as keywords (or phrases) carry their own meaning. Each entity class in Figs 4.3–4.4 and 4.6 has an entity type to provide definitional (internal) meaning.

Inference rules used for classification
A rule set is used to classify entities (hence associated objects) into classes according to attribute values for each entity identified. Thus, a rule set implementing a definition is used to test if an object (as a representation of an entity) belongs in a class. For example, a rule set implements the reason why a road or river entity object belongs to a specific road or river entity class as depicted in Fig. 4.3. It is also possible to develop classes for entities that are of special significance as determined through knowledge engineering interviews (Nickerson 1991, Ch. 3 this volume), map production specifications, and/or map design principles (Dent 1990).

The structural knowledge for classification is provided by functional dependencies and definitions. The structural knowledge for classification focuses on the internal meaning of classes as given by the bundle of attributes that are significant to each class. Structural knowledge for classification is generated and further processed by inference rules. Using the inference rules as process knowledge new classes are added, while others are eliminated from consideration. Entity classes for different phenomena can be grouped into associations by focusing on neighbourhood relationships.

Association

Domain role used for association
Associations are temporary neighbourhoods created on the basis of a significant relationship between entities. An association, as a loosely connected collection of entities, makes use of a domain role to clarify the nature of relationship affectation. In Fig. 4.5 with regard to the 'constrains' relationship between river path geometry and railroad bridge geometry, the river is playing the dominant role while the railroad bridge is playing a subservient role. Consequently, location of the river affects the location of

the railroad bridge. A role identifies which entity class provides the controlling influence of the relationship. Roles can also be developed for cartographic symbols to clarify significance of visual impression. In many cases visual impressions on a map are determined by which symbols are depicted as figure and which symbols are depicted as ground to preserve a visual hierarchy (Dent 1990).

Definition used for association

A definition provides the internal meaning of an association. The definition lists the entity classes that are involved in the association and provides a description of the relationship between the entity classes. The description includes what aspects of theme, space, and time participate in the relationship, and is referred to as a relationship-type definition. Locational displacement, as a relationship type between the river and road geometries of Fig. 4.5, is interpreted to be river dominant if the database emphasizes rivers, or road dominant if the database emphasizes roads.

Schema used for association

A prototypical spatial arrangement of features is defined using a schema. The spatial arrangement is important in characterizing the 'partially determines' relationship depicted in Fig. 4.5 between the highway path geometry and the highway bridge geometry, and between the railroad path geometry and the railroad bridge geometry. The relationships in this case are based on space rather than theme or time. An association does not have a requirement that the elements of the association have a part–whole role with one another as in the case of aggregation. However, an association is still able to imply a neighbourhood structure even though the relationships are not of a permanent nature as with aggregation.

Attached procedure used for association

Attached procedures compute associations that form the basis of neighbour-hoods, for example as given by the river and road association in Fig. 4.5. Particular conceptual dimensions of space, theme, and/or time underlie the relationships to be computed. Those objects closer in distance along the conceptual dimensions are more likely to be associated. This is the reason why a river and road that parallel each other have an association. One or the other usually dominates the locational displacement decision, depending on the significant character of the entities that is to be maintained.

Inference rule used for association

A rule set reasons about relationships among entities as in the case of Fig. 4.5 with the 'constrains' and 'determines' relationships. Other knowledge representation techniques such as schemata, domain roles, and attached procedures are used in the inference process to determine the kind of association, but the inference process directs what is appropriate. An inference process for associations deals mainly with entity to entity relationships rather than an entire set of entities at once.

An association is primarily a relationship among entities that is not easily named, but deals with spatial interaction. Consequently the results of processing, if retained, are retained only temporarily. What is retained permanently is how to intepret the relationships involved in structure recognition and structure identification once found. Domain roles and schemata are useful constructs for developing the structural knowledge of an association. Attached procedures are useful for updating a prototypical interpretation of what relationships are significant. Data in a particular association context are processed to update a default association. Inference rules guide whether each association is appropriate for each context in cartographic generalization. Entity classes and associations are represented at multiple levels of meaning with the assistance of an abstraction type called generalization.

Generalization

Type hierarchy used for generalization
A type hierarchy is the basis of a generalization hierarchy. The hierarchy is built over entity types using the primary spatial, thematic, and/or temporal keywords in the definition. The reverse process called specialization adds more specific characteristics to a class to 'specialize' a class. In Fig. 4.4, the class for bridge is specialized into railroad, highway, pipeline, and foot bridge; and subsequently highway bridge is specialized into interstate and arterial.

Definition used for generalization
Some spatial, thematic, and/or temporal aspects are more significant than others for entity classes. The most significant become part of the type definitions. An entity-type definition is created as a generalization from the definitions of subtypes. This implies a suppression of qualifying character-istics to make an entity class more general. An entity definition specializes into entity-subtype definitions by providing further qualifying characteristics. That process consists of adding necessary and sufficient characteristics to the definition to qualify the meaning.

Attached procedure for generalization
An attached procedure in a schema that implements a generalization hierarchy is used to find other references that possibly belong to the hierarchy. An attached procedure processes the validity of certain derived values for slots. They can be used to retrieve definition descriptions useful in the building of the generalization hierarchy. In Fig. 4.4 an attached procedure in a schema for 'bridge' would make sure that all relationships among entity classes (along with entity-type definitions) having some relationships to the concept 'bridge' are discovered and resolved.

Inference rule used for generalization

Manipulation of definitions is important to support the generalization–specialization process of Fig. 4.4. A rule set that will test the necessary and sufficient conditions to build a generalization hierarchy for the one in Fig. 4.4 requires technical details about bridge knowledge. Requiring such knowledge implies that knowledge about issues other than those of concern in the raw data are needed to solve map generalization problems. It is the knowledge on the bounds of a topic which ensures that what is known about a topic is well understood. Such knowledge can be acquired through expert sources in the fields of bridge design, hydrology, terrain analysis, and others required for a specific database.

Type hierarchies and definitions provide the structural knowledge for relationships among entity classes. The focus in generalization as an abstraction type is still on internal meaning of classes, dealing with the addition or suppression of significant characteristics in a type hierarchy. Process knowledge is represented by inference rules. The process knowledge controls the structural knowledge of general and specific meanings. The general and specific meanings are the foundation of the heterarchical processing for aggregating and disaggregating entities.

Aggregation

Domain role used for aggregation

In Fig. 4.6, the role of each localized terrain surface in the overall elevation surface is its contribution to the overall character of the surface. Role in this case consists of an internal (necessary) structural relation (Sayer 1984) in that the elevation surface is made up of those local surfaces, i.e. the elevation surface is what it is because of the parts of the surface, and would be something different if the parts are different. Consequently, each localized terrain surface determines the character of the more global elevation surface in a unique way. Rivers and streams each have their roles in a drainage network. The higher the order number of a watercourse branch, the more significant the role.

Definition used for aggregation

An aggregation requires a new definition once the aggregation is composed. That definition will be of a different character from the ones created through generalization. The meaning of the aggregation is a synergistic meaning due to a combination of elements, and perhaps different from just the sum of the other definitions. In the case of the localized terrain surface in Fig. 4.6, the definition is not substantially different. An aggregation of river and stream classes to compose a drainage network has a similar interpretation as the individual entity classes. If river and stream classes are combined with other entity classes the aggregation is likely to be different. An aggregation of river and road entity classes into a transportation network develops a significantly different definition from the individual definitions of river and road. An aggregation takes on a permanence of structurally related parts which is the fundamental difference between aggregation and association.

Schema used for aggregation

A schema is used to provide a default example (also called prototypical arrangement) of objects as for the highway and river classes in Fig. 4.3. A schema is used to represent the default characteristics of particular examples of highways and rivers. Structure signatures (Buttenfield 1991, Ch. 9 this volume; Buttenfield 1986, 1987) act as representative examples of entities in various phenomena domains, and provide a useful starting-point for processing. The several parameters of a structure signature are each included in a schema in the form of slot values.

Attached procedure used for aggregation

Attached procedures compute the nature of structural aggregations as part of structure recognition. Schemata with their attached procedures drive the recognition process. Such procedures support structure signature identification for all of the entities represented by schemata. Research on structure signatures for different types of phenomena will be useful in developing signatures for aggregations. Image schemata (Lakoff 1987) that characterize fundamental spatial relationships as part of human experiences, i.e. how we see things, might be useful in developing signatures for aggregations.

Inference rule used for aggregation

A rule set is used to reason about how to aggregate entities. Other knowledge representation techniques such as schemata, domain roles, and attached procedures are used in the inference process to determine the kind of aggregation that is appropriate. Inferences assist with aggregating the localized terrain surfaces of Fig. 4.6 into various neighbourhoods based on contiguity and the existence of other features. The inference process supports access to the surface at different levels of resolution depending on the neighbourhood context involved in a cartographic generalization.

Aggregations of geographical entities form geographical neighbourhoods when the fundamental building blocks are made of easily recognizable forms. Structural knowledge as definitions of entity classes and roles for entities are formulated using schemata as a best guess for the easily recognizable forms of such neighbourhoods. Process knowledge is represented using attached procedures and inference rules. Attached procedures are used to modify aggregations of features (entities represented by objects) in specific databases. Inference rules represent the reasoning why certain types of neighbourhoods are composed as they are. A neighbourhood commonly has a stable character in the world, and the same should exist in a database. When relationships are recognizable for neighbourhoods, but standard names are not useful in the long term for processing, then the temporary neighbourhood constitutes an association rather than an aggregation.

The abstraction types of Table 4.2 provide a basis for representing the meaning of individual entity classes as well as geographical neighbourhoods in an intelligent geographical database. The knowledge representation techni- ques in Table 4.2 are important building blocks for implementing the abstractions and are to be accessed by software during structure recognition and identification. Geographical neighbourhoods with source and target structures must be identified for submission to the cartographic generaliza- tion process. Structure recognition and identification in geographical neigh- bourhoods occur at multiple levels of conceptual resolution.

Geographical information abstractions implemented using the knowledge representation techniques are intended to characterize multiple levels of concepts. The problem does not appear to have easy solutions, as many techniques are required to represent sufficient knowledge. Whether or not the four abstraction types are the proper conceptualizations for the problem, and whether or not the knowledge representation techniques are appro- priate mechanisms, must be tested and evaluated either through more formalized treatment and/or through software implementations in a database environment. Although the formalized treatment seems appealing, the latter may actually provide a more interesting research experience. The next section briefly touches on this latter issue through a discussion of implementation considerations.

Implementation considerations in a database environment

It is possible to implement geographical information abstractions in various database environments. The author argues that bringing meaning on-line is essential to more effective results in digital map generalization. The rationale for this position is that too little information about meaning is the current approach, and this is contributing to the slow progress in digital map generalization. Consequently, three environments are discussed briefly as alternative implementation scenarios, including a metadata management environment as a separate data store, an embedded database environment with metadata required, and a hybrid of the first two.

A metadata management environment implements the information abstractions separate from the main data store. Such an environment preserves the performance of the original database, reducing the data volume that must be searched by conventional, spatial data processing operations for query and map production. Metadata management opera- tions that focus on deriving heterarchical information structures are run when geographical meaning for map generalization is required. Specialists in particular applications must be used to provide their perspectives on creating prototypic neighbourhood structures as part of metadata develop- ment. One disadvantage of the separate metadata is that complex software is required to maintain the relationships between the original data store and the metadata. Another disadvantage is that separate storage tends to encourage only periodic maintenance, resulting in a metadata set that easily evolves out of date.

An embedded database environment supports faster processing during map generalization since metadata are available immediately. Embedded metadata make maintenance of the metadata easier since they are close to the data store. When they are close to the data store they are more likely to be used. Data that are used often are kept up to date. Embedded metadata probably cause a slow-down in conventional spatial processing operations due to an increase in data volume. In addition, specialists are required to keep the information up to date. Specialists are difficult to find, and more expensive to retain.

Perhaps a better alternative is a hybrid environment, where metadata are processed as a separate store, but periodically loaded into the original database environment. Many database environments have periodic update cycles to maintain portions of a database in bulk. This approach takes advantage of both the other two environments without adversely impacting information systems operation, as long as storage structures for the original data do not depend on the metadata.

In all of the above implementation scenarios, an object-oriented database environment based on an object-oriented programming language makes an implementation easier to develop than does a third-generation software language (Parsaye *et al.* 1989). Although object-oriented database environments are still in a maturing stage and do not currently handle all abstraction types in native mode (Kim and Lochovsky 1989), they do seem effective enough to attempt a prototype system at this time (Mark 1991, Ch. 6 this volume).

Summarizing comments and future directions

Since the structure recognition problem is the first phase of map generalization, much of what follows that phase in terms of process recognition and the actual application of procedures depends on knowledge of geographical meaning. One of the main reasons why map generalization rules do not always succeed is because the process relies on context. The main component for understanding context is meaning. Developing geographical information abstractions in databases provides representation support for building intelligent databases. Geographical information abstractions are a way of harnessing more of the geographical reality intended in the data.

Both broad-based and specific research issues need to be addressed to make further progress in database/knowledge base representation to support map generalization. The broad-based issues concern the meaning of geographical meaning, and all ontological and epistemological issues necessary to formalize the concepts better. The specific issues concern the representation techniques to support the implementation of formalized concepts. Those issues have been and still are fundamental for geographical investigations. Pursuing such issues is important for making progress in map generalization in particular and information processing with a GIS in general.

As part of the broad-based issues for future research, conceptual issues underlying database abstraction and map generalization are in need of continued investigations. More studies clarifying the synergistic nature of database abstraction and map generalization would be useful. The big part of this problem is coming to grips with the use of a single term in two contexts, i.e. generalization. Although the author has tried for years to come up with a better term for both processes, the uses of the term are so entrenched in their respective fields of computer science and cartography that many more years are required to bring sufficient clarification. The dilemma remains unresolved as of this writing. However, we need to recognize the significance of both terms in their own context, and understand the different uses within those contexts, perhaps by using appropriate qualifiers. Such an activity will help extend our theoretical framework for database representations as well as map generalization. Clarification of the conceptual issues is a fundamental step in formulating sound techniques for representing geographical database meaning. Those sound techniques will be useful in developing better GIS tools that implement map generalization.

One solution for formalizing concepts is to have a good theory for the subject being modelled. At this point we are able to describe the geographical knowledge intended on a map using somewhat formalized high-level languages, which is better than trying to make natural language operational. Continuing to use a high-level, somewhat formalized language, that can be converted to a computational form, is probably the best way to proceed with examining the nature of the map generalization problem.

As part of the specific issues, a next step in this examination is to perform an assessment of the suitability of the representation techniques for database semantics. This examination will take some rigour and sufficient prototyping with respect to a few of the more significant techniques, for example definitions, schemata, and inference rules, as identified by Sowa (1984). More detailed examples need to be worked out, and prototype software developed to implement them. Object-oriented programming will be of considerable assistance since the constructs in object-oriented programs combine structures with operations in a single package (Mark 1991, Ch. 6 this volume). Such a combination provides for significant flexibility in the behaviour of the structural components of an object base, permitting exceptions to a rule as a matter of routine rather than as a matter of extra effort. The object-oriented programming approach is more akin to how humans qualify situations rather than like conventional programming languages where data structure is separate from operation, making behaviour of data difficult to change.

Sound concepts, techniques, and software tools are indispensable for modelling the nature of reality in a computer. Being able to capture and represent knowledge of meaning in geographical structure and process with such concepts, techniques, and software tools, will enhance not only studies in map generalization, but other studies generally concerned with geographical modelling. Geographical information abstractions as representations of meaning convey what has been missing, lost, or forgotten in many database models. Elucidating concepts, techniques, and tools for geographical information abstractions is an important step in developing more realistic digital geographical models.

5

Knowledge classification and organization

Marc P. Armstrong

Introduction

Although it is only part of the larger process of cartographic design, the problem of automating generalization is by itself a daunting one. Generalization has traditionally fallen partly into the artistic realm of cartography, in which the cartographer integrates knowledge about cartographic conventions and human cognitive processes in order to portray information in a way that conveys meaning to a map reader. These processes, however, are difficult to decompose into an atomistic set of facts and deterministic rules. Indeed, Robinson has stated that 'the artistic conception of cartography is *purposefully vague with respect to mapping rules* . . .' (Robinson *et al.* 1984: 14, emphasis added).

Generalization operations can also cause cascading effects which are difficult to treat with a rule-based approach. For example, if a rule is invoked to displace a feature, the displacement may affect the size and location of typography associated with it, which in turn, can cause other displacements to occur. To complicate matters further, generalization goals and outcomes are often ill-defined and thus generalization can be performed in many ways, with none demonstrably superior to all others. Because of these characteristics, cartographic generalization is a particular instance of a semi-structured problem that will not reduce to solution by a lock-step set of deterministic rules or a single algorithm.

Many semi-structured problems are attacked by ensuring that human decision-makers are empowered to act with an appropriate set of tools and decision aids (e.g. Armstrong *et al.* 1990). For generalization problems users of cartographic systems must combine their knowledge with that provided by the system to produce a map. Conventional cartographic software systems, however, have yet to provide an appropriate framework for representing, storing, and applying knowledge that can be applied to semi-structured generalization problems. The focus of this chapter is placed on designing a knowledge base that will support generalization decision-making in automated cartographic environments. The knowledge base as described contains categories of knowledge that must be applied in concert to accomplish generalization goals.

Background

There are two principal reasons why generalization is performed: to remove unwanted detail when a scale change takes place, and to remove unneeded detail for thematic mapping. Assuming that we start with a large-scale vector-mode cartographic representation of reality which contains much detail (e.g. Beard 1987), the problem then becomes one of removing superfluous detail and rearranging map elements to avoid conflict in the display as smaller-scaled output is required. Unfortunately, this seemingly simple process is complicated because it is poorly understood. Although we have a well-defined set of generalization operations (e.g. Brassel and Weibel 1988; Beard 1991, Ch. 7 this volume) in many instances, several procedures may need to be invoked to complete a generalization process (Brassel and Weibel 1988).

The interaction among these operations, and the specification of a logical sequence with which to implement generalization in a given context is unknown, however. McMaster and Monmonier (1989), for example, show how the set of raster-mode generalization operators that is invoked for a generalization operation changes with the scale of the desired product. Shea and McMaster (1989) also describe a series of generalization operations that can be implemented in some instances as a set of procedures based on geometrical algorithms; in such cases computational logic is embedded within the procedures. In other cases, generalization may require more sophisticated logic, and in such instances we can exploit facts and rules to implement generalization operations. Such declarative knowledge also can be used to guide the process of performing procedurally based operations, because in many instances, although we may know how to generalize, we may not know how much to generalize. The sequencing and parametrization of each step are clearly important tasks for automated generalization systems, and the cartographic knowledge base must be designed to support these processes.

Knowledge can take many forms and it can be stored and used in many ways. For this reason, a conscious effort must be made to structure the representation of knowledge so that it can be used to support map generalization decisions. Current knowledge-based systems, however, are best applied to tasks that are narrowly defined, clearly defined, and have outcomes that can be evaluated – the problem domain must be well structured. In fact, Holland (1986) asserts that rule-based systems are brittle, and are not adaptable to dynamic problem contexts. Additionally, a necessary prerequisite for developing rule-based systems is that we have available an expert, or better yet, a set of experts, from which we can extract knowledge (facts and rules) about the way in which problems are solved in an application domain. McMaster and Shea (1988), however, assert that decades of cognitive research will be required before we understand the rules by which human cartographers perform generalization operations.

Though we must clear many hurdles before producing fully functional automated generalization systems, we must begin to devise intelligent strategies to free users from mundane technical details of generalization and to assist users in cartographic production tasks. Generalization normally fulfils a set of philosophical, application, and computational objectives (McMaster and Shea 1988), that must be elicited prior to the start of generalization. These objectives are translated into an operational generalization strategy by drawing upon knowledge about the problem domain, and solution strategies for solving problems in particular contexts. This knowledge about the development of strategies is referred to as meta-knowledge (Wah and Li 1989), and the process of applying it in particular applications to develop scenarios is known as metaplanning. Meta-knowledge, therefore, is used with elicited generalization goals by the metaplanner to provide a workable strategy for invoking generalization operations and for providing support for cartographic decisions.

It is important to begin to think of ways to operationalize existing conceptual frameworks for generalization, and to categorize knowledge that will enable the problem of automated generalization to be systematically attacked in stages. This incremental development of a knowledge-based generalization strategy to support cartographic decisions is consistent with Weibel's view of generalization described in Chapter 10 (Weibel 1991). One approach to treating the complexity of a map generalization knowledge base is to decompose the problem of knowledge representation by forming a set of categories to facilitate the collection, management, and application of knowledge about generalization. In the remaining sections of the chapter a basis for classifying knowledge needed to automate generalization is described; a general strategy for evaluating and using the stored knowledge is also briefly discussed.

Knowledge classification and representation

Databases contain representations of observations about the real world. Such observations are converted to meaningful information when they are placed in the human context, and when they provide material that was previously unknown, or inaccessible to a user (Wiederhold 1986). Knowledge, and therefore a knowledge base, is concerned with a more abstract generalization of material that comes about from education and experiences that occur as humans are confronted with problems. In many application domains, there is a move away from record-keeping Data Base Management System (DBMS) functions, to those which require access to a variety of different information and knowledge types (Dayal and Smith 1986; Manola and Brodie 1986). A knowledge base, however, must contain schemes that allow access to both information and knowledge. Other knowledge resides with users, and interaction and control structures enable decision-makers to bring their personal knowledge and expertise to bear on problems.

Three kinds of stored knowledge are necessary for effectively implementing rule-based systems designed to perform cartographic generalization: **geometrical**, **structural**, and **procedural**. **Geometrical knowledge** is similar to what cartographers would recognize as a set of feature descriptions encompassing absolute and relative locations, but it includes other aspects (e.g. feature density) important to generalization as well. **Structural knowledge** brings expertise that ordinarily resides with the cartographer into the automated generalization process. Such expertise, for example, may be derived from geomorphology, hydrology, or cultural geography, and may be conditioned by past experiences and practices adopted by mapping agencies. Finally, **procedural knowledge** is used to select appropriate operators for performing generalization tasks. These three kinds of knowledge must be represented in a coherent framework that allows declarative and procedural knowledge to interact as it is evaluated.

Knowledge can be described using several strategies such as object–attribute–value (OAV) triplets, production rules, and semantic nets. It can be argued that OAV triplets, frames, and semantic nets exist along a continuum of knowledge aggregation, with OAV triplets representing elemental portions of knowledge which can be aggregated into frames, which, in turn, can be linked into a semantic network (Harmon and King 1985). In general, as increasingly structured strategies (e.g. frames) are used, inferencing times decrease and memory requirements increase because pointers are often used to implement structured knowledge representation schemes (Wah and Li 1989).

Frames are used here as a tool to illustrate knowledge about cartographic generalization. Frames are useful when both data and knowledge must be represented for an application domain (Fikes and Kehler 1985). Frames permitting the specification of both declarative and procedural knowledge (Minsky 1975; Parsaye *et al.* 1989: 183–92) have been cited (Nyerges 1991c) as useful for representing knowledge for a generalization task (classification), and have been used to represent knowledge in several other spatial applications (e.g. Kuipers 1978; Smith, Pellegrino, and Golledge 1982; Armstrong *et al.* 1990). Frames are also useful analogues for objects (Parsaye *et al.* 1989) and object-oriented approaches to generalization are advocated in the literature (Mark 1991, Ch. 6 this volume; Nyerges 1991c; Armstrong and Bennett 1990b). The translation from frames to object-oriented environments is straightforward. The intent of this chapter, however, is not to argue that there is a single best method of knowledge representation, but rather to illustrate how knowledge can be represented to improve the process of cartographic generalization. The debate about knowledge representation is likely to continue for some time, and the resolution of the debate lies in the evaluation of fully implemented systems.

Each frame has a name, and a list of slots or facets. Each facet contains either a value with its associated descriptor (in measurement units), or a procedure that is executed to produce a value (Fikes and Kehler 1985). This latter capability is sometimes referred to as procedural attachment. Thus, an automobile could be described as a series of slots, associated values and labels (Make: Jaguar; Year: 1985; Colour: British Racing Green; Original_cost: £19 500; Sale_price: £12 999). A method for determining monthly finance costs could also be included as part of the frame. Also note that other knowledge about an automobile, such as its maintenance history, could be stored in other frames.

As used here, a frame has the following generic structure (De 1988):

```
(frame
    (slot ( facet (datum (label message . . .))
                  (datum . . .) . . .)
          (facet . . .) . . .)
    (slot . . .)
. . .)
```
[5.1]

This frame-based approach to knowledge representation also allows rules to be specified. Rules are encoded using the following generic structural form:

```
(Rule Example
    (AKO ($Value (Preprocessing_Rule)))
    (Precondition ($Value (< variable value)))
    (Consequent ($Value (Do_Something))))
```
[5.2]

The frame has a precondition that consists of an arithmetic operator, a variable, and a value of that variable. If this precondition evaluates to true then the consequent (Do_Something) occurs. The hierarchical structure of a frame is a rich mechanism for declarative knowledge representation because many spatial relationships fall naturally into this form, and where they do not, the structure can be expanded to represent knowledge by linking additional frames into a network. In this case, the structure is given by identifying the rule as a kind of (AKO) preprocessing rule.

Geometrical knowledge

The objects represented in this section are based on the proposed standard for digital cartographic data (NCDCDS 1988). Within this standard, members of a class are referred to as features, and consist of a real world phenomenon (entity) and its digital description (object). Table 5.1 shows

Table 5.1 Selected cartographic objects defined in the proposed Digital Cartographic Data Standard (DCDS)

	Geometry	Geometry and topology
0-D	Point	Node
1-D	Line segment	Link
	Arc	Chain
	Ring (string)	Ring (chain)
2D	Polygon	Polygon
	Pixel/grid cell	

selected classes that are often used in cartographic representations.

The following frames represent extended versions of selected NCDCS vector-based primitive spatial objects:

(node
      (AKO ($value (point)))
      (ID ($value (pointid)))
      (ATTRIBUTE ($value ((node_attributes))))   [5.3]
      (INDEGREE($value (number_of_links)))
      (CONNECTED_TO ($value ((chainids)))))

(chain
      (AKO ($value (feature)))
      (ID ($value (chainid)))
      (ATTRIBUTE ($value (chain_attribute)))
      (FROM_NODE ($value (nodeid)))   [5.4]
      (TO NODE ($value (nodeid)))
      (LEFT_POLY ($value (polyid_list)))
      (RIGHT_POLY ($value (polyid_list)))
      (BUILT_FROM ($value (point_list)))
      (DIST_IF_NEEDED (method to calculate chain length)))

Note that most implementations of chain objects provide for a single left and right identifier. This representation is necessary but not sufficient to support all generalization operations. Consider, for example, when a chain represents a river and a border between two political jurisdictions. In such cases decisions to remove chains, say on the basis of stream order, should not disrupt the political boundary between areas. In addition, borders are often hierarchical. In the USA, for example, a chain may represent not only a river, but also a county and state boundary. Such problems can be treated by employing a hierarchical attribute code similar to that used by USGS digital line graphs (USGS 1985), or by implementing a list of identifiers as illustrated in frame 5.4.

Because the way in which an object interacts with its neighbours is a key determinant of its local importance, topological relationships are required to support generalization decisions. Topological relationships can be calculated using methods or, if they are frequently required for an application, they can be stored in a frame. In the following example, stream order is encoded:

(Stream Topology
      (AKO ($value (topological_descriptor)))
      (ID ($value (stream_id)))   [5.5]
      (Stream_order ($value (level))))

The use of methods and demons (e.g. DIST_IF_NEEDED) extends geometrical knowledge beyond a simple set of descriptions of cartographic objects. In frame four, for example, a method would be invoked to calculate distances if they are required in the course of making a generalization assessment and other methods could be included to compute centroids or areas of objects. Methods may also be required to generate buffers which enable knowledge about congestion in a neighbourhood to be calculated by enumerating objects within the buffer zone. In such cases, additional knowledge must be encoded to facilitate decision-making about the intersection of a buffer and objects. The size of the buffer, of course, depends on the scale and purpose of the map, which are among the goals that must be elicited at the outset of the map creation process (e.g. Armstrong and Bennett 1990b).

Structural knowledge

Cartographic features result from numerous generating processes, some natural (e.g. vulcanism), others human (e.g. street grid), and some which fall in an intermediate realm (e.g. channelized streams). Structural knowledge arises from the generating process of an object, and is used to provide guidance in automated generalization because the generating process determines, in part, the way in which a feature is depicted. Jenks (1989) asserts that features created by humans tend to follow smooth curves and have straight-line segments, while features arising from natural processes tend to be more complex. For example, a railroad has a different generating process from a river, although both are one-dimensional objects at most display scales. The generating process in this example, however, has a decided impact on the sinuosity and line character of the object types. Thus, the application of structural knowledge aids in the selection of appropriate generalization operators and in parameter specification because the context in which generalization takes place is better defined.

Consider, for example, the following frame:

(line_type
      (AIO ($value (chain)))
      (ID ($value (90)))   [5.6]
      (SCALE_ENCODED ($value (24 000(:unit ratio))))
      (Feature_type ($value (rugged_coastline))))

In this example, a line (ID = 90) is identified as an instance of (AIO) a chain. The feature represented by the chain is a rugged coastline, as determined by an expert geomorphical evaluation. This kind of structural knowledge can also be used later in the generalization process, for example, to guide the selection of a smoothing operator. Note that this structural knowledge is used to supplement the logic that is embedded within the code used to perform simplification operations.

This approach represents a formal yet data-intensive specification of knowledge, and the implementation of such a scheme would require a large effort to collect, encode, and efficiently organize it. One way to reduce knowledge elicitation and encoding steps is to use existing information that is often stored in GIS layers to support the acquisition of structural knowledge. Consider an example in which the location of faults and materials of different resistance to erosion is stored. The geomorphic process which gives rise to the form of a stream can provide insight into how the feature should be depicted at different scales.

Geological control, for example, influences the basic drainage pattern of a region (e.g. rectangular or parallel). Geological control also affects drainage density which is the average length of streams per unit area. Ritter (1986: 168) reports that density varies from 3–4 miles (4.8–6.4 km) per square mile in an area with resistant sandstone to 1100–1300 miles (1770–2092 km) per square mile in an area characterized by weak clays. Knowledge about such basin characteristics can be applied to simplification tasks. In addition to basin characteristics, individual streams are also subject to geological control. Channels, for example, are often placed into straight, meandering, or braided classes. These general forms are related to slope, climate, and geological factors (Gregory and Walling 1973), and they can be measured and stored as structural knowledge to aid in the selection of appropriate parameters for generalization.

Furthermore, Leopold, Wolman, and Miller (1964: 297) posit that meander amplitude is controlled more by stream bank material than by hydrodynamic factors. In either case the geometric characteristics of a generalized chain should be influenced by structural knowledge about controlling factors. Other structural knowledge comes from computations performed on the existing cartographic database (e.g. sinuosity, Buttenfield 1989, and length, Buttenfield 1991, Ch. 9 this volume). These measures may be derived (using procedural attachment) and retained during the process of making a decision about generalization.

Procedural knowledge

Procedural knowledge is used to guide the selection of appropriate generalization operators and algorithms in a given map context. This may involve a decision about whether a feature must be modified through simplification, or whether it should be displaced to relieve congestion and improve legibility (Monmonier 1989b; Armstrong and Lolonis 1989).

McMaster (1986) describes measures that can be used to evaluate simplification and data reduction algorithms. These measures also can be used to formulate rules which help select, using empirical evidence, an algorithm for a particular task. For example, McMaster states that the percentage change in the number of curvilinear segments is a good indicator of how a smoothing operator affects a line in comparison to a measure of percentage change in angularity. The percentage change in angularity, in turn, evaluates reductions in small crenulations after a line is simplified. These and other measures can be used to evaluate different algorithms for a specific task.

In some instances an operation that eliminates all small crenulations may be required, whereas in others, some small crenulations may be desirable to give the line character. White (1985) reports that when generalization algorithms are compared to human selection of critical points, the Douglas–Peucker algorithm performed best of those tested. In other studies conducted by McMaster (1987) and Jenks (1985, reported in McMaster 1987) the Douglas–Peucker algorithm also fared well. Additional research is needed to determine the most appropriate algorithm for specific geographically based generalization tasks, or to determine whether a single algorithm is best suited for all tasks. If such knowledge becomes available, then it should be incorporated into the knowledge base to increase the effectiveness of generalization processes.

Procedural knowledge also can be used to guide the placement of typography on maps. Several successful implementations of using rule-based approaches to place-names associated with point symbols have been reported (e.g. Jones 1989). The rules in these applications search for free space, avoid overlap, and position names about the point features in a hierarchical order, with right and top usually taking precedence. Although these rule-based approaches are useful for assigning labels, little work has been performed on the process of determining which features to include on the map. In one notable exception, Langran and Poiker (1986) describe several procedurally based approaches to determining whether settlements should be retained or deleted as one part of a point labelling process.

Using knowledge in generalization

Assuming that an appropriately specified knowledge base can be created, the next problem arises from its use. This implies the existence of a conceptual framework on which to base the application of the knowledge base. Such a framework can be constructed from strategies developed by Robinson *et al.* (1984), McMaster and Shea (1988), Shea and McMaster (1989) and McMaster (1989b). In each of these frameworks, selection is treated as a preprocessing step which is followed by a series of generalization operations that are dependent upon the purpose of a map and its scale.

Selection

Selection refers to the process of deciding which classes of information, and perhaps which objects within a class, are needed for the purpose of a map. Robinson *et al.* (1984: 125) state that the selection of objects to be mapped is not part of the process of generalization. Because selection provides for the establishment of the set of objects which are no longer needed for a map, and which therefore, will not be subjected to further processing, it may be useful to ensure that selection be considered as part of the generalization process in the digital cartographic realm. In practice, the selection process involves a series of dichotomous decisions, in which no modification of an object takes place (McMaster 1987).

The task of selection is difficult to divorce from other related generalization operations, because selection may depend upon scale, feature density, and symbolization. The selection of objects for a given display is also related to the theme of the map and its intended audience. If, for example, a cartographer is preparing a map depicting a macro-scaled biological phenomenon for New York State, then including all primary and secondary roads in the state would possibly lead to excessive clutter. On the other hand, if the map were intended to show impacts attributed to the application of road salt, then roads may be required to provide an appropriate frame of reference. Selection is exceedingly difficult to automate. At present it is best left as part of the interactive decision-making process. In fact, the specification of map objectives is an important part of the process of metaplanning that is used to establish the way in which other generalization tasks are structured.

Simplification

Simplification refers to the process of eliminating unneeded detail from a map, and geometrical, structural, and procedural knowledge can be profitably applied to the problem. Three main types of simplification (Muehrcke 1986) are considered in this section: point simplification, line simplification, and feature elimination.

To elaborate on point simplification, assume that a homogeneous set of point symbols is used to represent the distribution of individual objects on a detailed map. Simplification enables several features to be represented by a single symbol on a map produced at a smaller scale; the distribution, or clusters of symbols within the distribution, are typified by a single symbol taken from the original set (Robinson *et al.* 1984: 252). If quantitative data are used, then the weight associated with a point may be used to aid in making a decision about symbol retention. If the data are nominal scaled (e.g. oil wells in an oilfield) then the process becomes more purely geometrical in the sense that each object in a class has a weight of one and decisions are made on the basis of location. One way to structure this problem is to use an algorithm to make a preliminary assessment of the distribution, and then to use rules for evaluating the results of the algorithm.

The *p*-median model has been used extensively in locational analysis to determine sites for the location of facilities (Hillsman 1984), and can be readily adapted to the process of cartographic generalization when features are located on a network (e.g. settlements). Several ways of solving the model have been developed, although a heuristic method (Teitz and Bart 1968) has been widely adopted. In the model, *p* represents the number of sites to be included in the final solution. Nodes are selected with the objective of minimizing average distance between each node and the nodes that are closer to that node than any other node. As it is applied to cartographic generalization, each feature is a candidate, the value of *p* can be specified on the basis of the desired scale of the map, and a set of rules can be formulated to specify appropriate ranges of scales and items displayed.

(elimination_rule 12   [5.7]
    (AKO ($value (elimination_rule)))
    (PRECONDITION ($value (= elimination simple_point)))
    (CONSEQUENT ($value (compute_*p*-median))))

In the above frame, a rule is given to invoke a method to compute a *p*-median solution to a set of points identified as candidates for point simplification. Note, however, that the *p*-median solution is computationally intensive and has several data requirements which must be satisfied (e.g. a distance matrix must be computed). In addition, the number of sites to be located, or in this case, the number of points to be displayed, must be specified prior to the execution of the process. This can be accomplished through the use of rules:

(*p*_selection rule 10   [5.8]
    (AKO ($value (elimination rule)))
    (PRECONDITION ($value ((> scale_factor 50 000) and
                            (< scale_factor 100 000))))
    (CONSEQUENT ($value (*p*_% = 15))))

In this example, 15 per cent of the original points would be retained for a map produced at scales between 1 : 50 000 and 1 : 100 000. This percentage value would need to be translated into an absolute number for the algorithm to proceed. Clearly, the formulation of robust rules will be a difficult task, although rule formulation can be guided by existing cartographic practice and knowledge. For example, a method for determining the reduction in numbers of items that will occur when changing scales (e.g. Töpfer and Pillewizer 1966) can be used.

The second type of simplification focuses on linear features. Line simplification is perhaps the most studied form of cartographic generalization (see e.g. McMaster 1987). In most instances the process of line simplification relates to the application of an algorithm to reduce the number of points in a line while retaining its original form or caricature. In a rule-based system the main task lies in the selection of an operator that will perform best for a given scale and map purpose. In this instance, procedural knowledge is paramount, although other knowledge can be used to increase the effectiveness of any purely geometry-based algorithmic approach to generalization.

```
(smooth_choice                                          [5.9]
        (AKO ($value (smooth_operator)))
        (Feature_type ($value (rugged_coastline)))
        (SCALE ($value (S)))
        (Recommended_smooth ($value (Douglas_Peucker)))
        (TOLERANCE ($value (X))))
```

In this example, the Douglas_Peucker algorithm would be selected as the simplification algorithm for rugged coastal environments.

Feature elimination is the third type of simplification. For feature elimination, a decision must be made about whether to display an object, given the purpose and scale of the intended map. Two major criteria can be used to provide guidance about feature retention: geometry and attributes. Monmonier (1986) has suggested that stream geometry be included in a digital database so that only streams with a certain stream order would be shown at a given scale. This can be accomplished by feature tagging, or through the use of a transfer function. Features in such a tagging scheme would be eligible for inclusion in a display only within a range of scales. This ignores, of course, the problem of symbolization (e.g. cased lines for a stream), but feature retention is a necessary prerequisite step. Assume in the following rule that the target scale of the map has been determined:

```
(Rule 52                                                [5.10]
        (AKO ($value (Simplification_Rule)))
        (Precondition ($value ( < stream_order 2)))
        (Consequent ($value (Do_not_add_to_plot_list))))
```

This rule is a simplification rule in which a precondition is tested for compliance (is the stream order less than 2?) and a consequential action to be taken if it evaluates to true is specified (do not plot the stream). The precondition can be any kind of arithmetic or Boolean expression. A fact similar to that which would be used in evaluating the rule can be found in frame 5.5.

Monmonier (1986) also described a strategy in which points are flagged in an attempt to emulate how a cartographer trained in geomorphology would simplify a fjorded coast while retaining its distinctive geomorphic character.

```
(saliency                                                    [5.11]
    (AKO ($value (point_list)))
    (ID ($value (90)))
    (list ($value ((salient_point_list)))))
```

In this example, a list of points with geomorphic saliency is given. These points would then not be candidates for elimination by simplification algorithms. Feature tagging can also be implemented hierarchically so that saliency is scale dependent. Thus, the salient point list would change as restricted by a scale parameter.

Tagging also can be related to attributes such as those used by the USGS for their feature codes for roads and trails in digital line graph (DLG) databases (USGS 1985). The attribute code identifies each chain by a description based on road size and trafficability (e.g. primary route, class 1 divided, lanes separated). These feature codes can be used to control the volume of data used in an analysis, or they can be used to control which features will be used in a display.

```
(Plot_rule 90                                                [5.12]
    (AKO ($Value ( Simplification_rule)))
    (Precondition ($value ((= road_class 4) and (> scale factor
    125 000))))
    (consequent ($value (do_not_add_to_plot_list))))
```

Additional structural information may also be important in determining whether features should be included on a map. Gardiner (1982) suggests that geomorphical and hydrological characteristics can be used in the process of identifying streams for plotting on a map. In his example, parameters such as channel width and discharge are used. In this way it is possible to capture the richness of knowledge than an expert cartographer would bring to the generalization process.

```
(Geometrical_fact 23                                         [5.13]
    (AKO ($Value (Simplification_fact)))
    (OBJECT_ID ($Value (90)))
    (CHANNEL_WIDTH ($Value (30(:unit metres))))
    (DISCHARGE ($VALUE (1200(:unit CMS)))))
```

```
(Structural_rule 7                                           [5.14]
    (AKO ($Value (Simplification_rule)))
    (PRECONDITION ($Value (( < CHANNEL_WIDTH 20)
                          and
                          (< DISCHARGE 200) and
                          (< MAP_SCALE 125 000))))
    (CONSEQUENT ($Value ( Do_not_add_to_plot_list))))
```

In this example, facts about each object are stored in the knowledge base as geometrical knowledge, and structural knowledge is used and applied to the facts to determine suitable plotting strategies.

Note that the rules used to make a decision about whether an object should be displayed are complex and may be mediated by the number and arrangment of features in a region of the map (for example, see any map of Australia in a world atlas). This introduces considerable complexity into the process, and may require a conditional selection, and then a subsequent resolution of conflict through the use of primacy when conflicts occur. This potentially involves large amounts of iterative calculation which becomes even more demanding if irregularly shaped objects are considered. Object-oriented programming approaches (Mark 1991, Ch. 6 this volume; Armstrong and Bennett 1990b) may simplify this process because objects contain a description of their characteristics and methods can be developed so their size is known at a given scale. Smalltalk V has built-in methods to determine if rectangles overlap, and this capability can be used to assess overlap among the extents (bounding rectangles) of objects. Object-oriented programs work by sending messages between objects, and thus neighbour-hood operations and the determination of local congestion in irregularly shaped regions or buffers can be handled.

Classification

Robinson *et al.* (1984) describe two types of processes which fall into the category of classification. The first involves assigning similar phenomena to categories. In this case it is primarily the attributes of features that are used to provide the basis for the classification. Because attributes are used, their level of measurement is important in formulating the classification. A qualitative classing can occur in two ways: hierarchically and by expert judgement. A good example of a hierarchical classification scheme for qualitative data is found in Anderson *et al.* (1976). In their scheme land use and land cover classes are increasingly resolved (e.g. forest land at level 1, and deciduous, evergreen, and mixed at level 2), thus enabling the use of the scheme when sensors with different levels of spatial or radiometric resolution are used. A non-hierarchical classification exists when there is a logical basis for grouping objects. The specification of an explicit mapping (assignment) of features into classes is required to implement this approach. When quantitative attributes are used in a classification, it is often based on measures of statistical similarity (e.g. clustering algorithms).

Whereas simplification is applied to objects to determine which should be retained for display, the second major process of classification involves selecting a location that typifies the underlying distribution of features (Robinson *et al.* 1984). This process may require the display of a symbol, for example, at the centre of gravity of a cluster of features. If the purpose of the classification is geometrical, when points are considered, they can be subjected to algorithmic analyses by using a two-dimensional clustering algorithm (Scott 1971: 45). This heuristic algorithm finds the median points (m) of a distribution, where m is a user-supplied parameter. This process operates in much the same way as the *p*-median approach described above; it differs in that the result need not be located at a defined location of an existing feature.

Knowledge organization and evaluation

How should the facts and rules contained in the knowledge base be evaluated? Although rule-based systems may employ several strategies for evaluating the knowledge and data needed to arrive at a decision, most systems employ first-order predicate calculus in which *modus ponens* reasoning is used. This provides a system of logical deduction that is the basis of the PROLOG language (Sterling and Shapiro 1986). Many rule-based systems employ PROLOG, and are often referred to as production systems in which rules consist of a condition (IF this) and a consequent (THEN do that). In the process of evaluating rules in a production system, if more than one rule has its condition satisfied (the IF part is true) then a conflict (which rule is fired?) must be resolved through a priority ordering on the rules (Holland *et al.* 1986: 14); these systems, therefore, often contain a central algorithm that chooses one rule to be fired. In cartographic generalization, and other semi-structured problems, such ordering schemes are difficult to develop and defend. As a result of these problems, researchers have begun to explore alternative ways of dealing with actions to be taken after rules are evaluated.

Holland *et al.* (1986) argue that semi-structured problems must be treated by using mechanisms for augmenting structured ways of analysing problems. One approach is based on the use of inductive logic and message passing (Holland 1986; Holland *et al.* 1986; Frey 1986; Booker, Goldberg, and Holland 1989; Armstrong and Bennett 1990a). Holland *et al.* (1986) state that the IF–THEN structure of rules in a production system is rigid, and maintain that whether a rule is executed should depend upon how it fares in competition with other rules. The hallmark of this approach is that when rules are evaluated, they are not treated as a command. Rather, a 'flock' of rules may be offered up and put into competition as evidence about what is transpiring in the system is uncovered. In this inductive framework, rules are treated more like suggestions than commands when they are evaluated in the affirmative. In such modes, rules are evaluated to provide evidence about hypotheses related to the problem domain, and as evidence is collected from several general sources it is then related to specific problem instances. Rules therefore act as building blocks of knowledge about the system. As evidence accumulates new structures are constructed using genetic algorithms (Goldberg 1989), and are evaluated inductively, until ultimately actions are taken.

The approach used by Holland *et al.* (1986) to implement their ideas involves the use of bit-string classifiers which can be constructed readily from a knowledge base organized using frames. Holland *et al.* (1986) also allow for parallelism in the rule evaluation process. Parallel processing will provide a needed increase in the performance of cartographic generalization operations (see Langran 1991, Ch. 12 this volume). Holland *et al.* (1986) also suggest that limited parallelism can enable several rules to be fired simultaneously, thus forming a network of interacting, competing hypotheses. This can help to form nascent learning through rule association. The approach is similar to a connectionist approach, but it differs because the messages can be complex, and such complexity violates the connectionist principle that each processor cell provide limited memory (Hwang, Chowkwanyan, and Ghosh 1989). Improvement in performance attributable to parallel processing is especially important if human designers are interacting with the system during the map production process. It remains to be seen, however, whether this parallel, inductive approach can be fully implemented in cartographic generalization applications.

Summary

The geometrical and topological structure of conventional cartographic databases is well developed, but other information and knowledge that can be used to support generalization tasks normally are absent. Existing databases must be augmented to support generalization decision-making because many cartographic generalization problems remain analytically intractable, and must be attacked through the application of rules to existing or derived facts. Three kinds of knowledge provide separate, yet complementary, assistance to the process. Geometrical knowledge is used to describe features and their relationships, and can also be used in characterizing objects for further processing and evaluation. Structural knowledge is used to describe the underlying control of the digital cartographic representation of an object. Structural control influences how an object is treated by generalization operators. Finally, procedural knowledge is used to select a particular type of operator and to perform logical operations on the digital representation of cartographic objects. This knowledge-based approach is superior to storing multiple copies of a database, or to reliance on purely analytical approaches to generalization.

Important areas of future research will centre on how existing information stored in GIS databases can be transformed into knowledge to support generalization decisions. This research is especially needed because of the volume and complexity of knowledge that is needed to implement systems designed to support generalization. Despite the existence of a rudimentary implementation (Armstrong and Bennett 1990b), further work must also be performed on the development of a framework to support the acquisition of metaknowledge about generalization, and on the implementation of a fully functioning metaplanner. This research will provide a needed overarching control structure that applies existing facts and rules contained in generalization knowledge bases to cartographic problems. Finally, because the process of evaluating knowledge in many existing systems is inappropriate for supporting decisions about semi-structured problems, alternative evaluation strategies must be developed. Induction, supported by bit-mapped classifiers and genetic algorithms, should be explored as feasible mechanisms for evaluating rules and facts applied to semi-structured cartographic generalization problems.

Acknowledgements

Partial support for the original draft of this chapter was provided by the Center for Advanced Study at the University of Iowa; special thanks to Jay Semel and Lorna Olson. The chapter benefited from the synergism that took place among the participants at the Syracuse Symposium. The anonymous reviewers also contributed critiques which improved the chapter. Thanks also to David Bennett and Frank Weirich for intermittent conversations about implementation issues and process geomorphology.

6

Object modelling and phenomenon-based generalization

David M. Mark

Introduction

Generalization is an important process in spatial data handling. Not only is it needed for graphic display, but it may also be important for efficient and appropriate spatial analysis (Brassel and Weibel 1987). Furthermore it has long been recognized that cartographic generalization must pay attention to geographical factors as well as those of graphic design. Pannekoek (1962) emphasized that the generalization of coastlines and contours is at least in part an exercise in 'applied geography'. However, when computers were first applied to the line generalization process, attention was (understandably) given first to a geometric approach to the problem, based on algorithmic, procedural methods. Recently, it has been suggested that a more geographical approach to generalization should be adopted (Mark 1989).

In order to support phenomenon-based ('geographical') generalization in cartography, an expert systems approach, using rules, will be appropriate. Adopting an object-oriented approach to representation and software development might enhance the implementation of such concepts. Specifically, four high-level classes (hyperclasses) of objects are needed: **entities**, which exist in the geographical 'real' world; **symbols**, which exist in the cartographic world; **objects**, which exist in the digital world; and **features**, which are composed of the entities, symbols, and objects with the same real-world referents. Many geographical or cartographic features will have representations in all four of these hyperclasses, and thus this is a problem of multiple representation (see Buttenfield and DeLotto 1989). If the classes are properly linked, then generalization rules may be applied in one hyperclass domain, and the effects will propagate through to the others. For example, the Douglas–Peucker algorithm (Douglas and Peucker 1973) deals with chains (polylines) and strips (bands) in the digital object domain, with the results transferred to the cartographic symbol domain only when the line is plotted. Since generalization rules and principles in the digital object domain are fairly well known (see McMaster 1987 for a detailed review), this chapter concentrates on object-class definitions and rules for the symbol and entity domains.

The second part of the chapter presents a collection of instructions (rules) for compilation of 1 : 24 000 scale US Geological Survey (USGS) topographic maps from aerial photographs and field surveys. Although compilation and generalization are different processes, the rules for producing a map of 1 : 24 000 scale from cartographic data at much larger (cadastral) scales would probably be quite similar. Attention then focuses on the possible implementation of these guidelines in an object-oriented, rule-based expert system, and on research needed to rationalize geometrical thresholds inherent in many of the rules.

Object-oriented methods

Object-oriented programming and associated database methods represent an important and relatively new approach with considerable potential in spatial data-handling. The object-oriented approach emphasizes the definition of classes for objects, relations between classes, and possible operations on classes and their members. Progress in cartographic generalization will be achieved by attempting to model and generalize real-world objects or features, rather than their cartographic representations (Mark 1989). It would appear that an object-oriented approach will allow such methods to be implemented and tested.

A central concern in the object-oriented approach is the identification of specific object classes to be represented. In particular, the concern is with finding classes of objects with common behaviour: 'Perhaps the most useful technique for finding classes . . . is to look for meaningful external objects. Many classes just describe the behavior of objects from the abstract or concrete reality being modeled – missiles and radars, books and authors, figures and polygons, windows and mice, cars and drivers' (Meyer 1988: 326). Notice that these object classes can be almost anything – Meyer's examples include entities from the 'real world', graphic or digital representations of that world, hardware peripherals, and data structures. Classes are often arranged hierarchically, and inheritance, especially multiple and repeated inheritance, is another central concept.

Of course, as is the case with many new computing approaches, object-oriented approaches have been misused, misunderstood, and oversold (see especially the discussion in King's 1989 paper, 'My cat is object-oriented'). Nevertheless, the approach does have appropriate roles within geographical computing, and this chapter reviews a promising role. For surveys of object-oriented approaches, see Meyer (1988) or Nierstrasz (1989). For an introduction in a geographical or GIS context, see Duecker and Kjerne (1987), or Egenhofer and Frank (1989). Hu (1989) has suggested that C++ can provide a useful environment for applications combining an object-oriented approach and an expert systems framework.

Object definitions for cartographic generalization

As noted in the introduction to this chapter, the approach proposed here requires the identification of three high-level classes of objects, including entities, cartographic symbols, and digital objects. This section reviews each in turn.

Entities

One major object class would include features of the real world; defining the object classes here would be an exercise in geography or its subdisciplines (geomorphology, for example), and these are termed 'entities' in the NCDCDS (1988) terminology. This class would have subclasses for things such as shorelines, settlements, roads, lakes, railways, rivers, etc. Within shorelines, subclasses would include landform primitives such as spits, stacks, fjords, bars, tombolos, etc., and also ensembles of landforms, such as barrier-beach coasts and ria coasts. Attention to the problem of representing geographical information is also a theme in the work of Nyerges (1991c) and of Armstrong (1991) (Chs 4 and 5 this volume).

Cartographic symbols

The second major class would include cartographic symbols. These are conventionally divided into points, lines, and areas, but could also include graphic primitives for raster images. Properties of such objects are largely graphic, and include point symbol sizes and shapes; line widths, patterns, and textures; areal fill patterns; colours for any of these, etc. For topographic maps, categories ('layers') such as hydrography, contours, transportation, survey, etc., form the next higher 'level' of classes; the feature-coding systems of USGS digital line graph (DLG) and Geographic Names Information System (GNIS) data, or similar data from the US Defense Mapping Agency (DMA) and other mapping agencies, would form the core of the class definitions here. Note that these classes are in some sense geographical and in other senses cartographic.

Road maps would have a different symbol dictionary, with some properties in common across almost all maps, others common across just one country's maps, still others related to a single publisher or map producer. A rule base for these cartographic symbols could involve simply an inventory of the present practice of the publisher or cartographer that the computer system is to serve or emulate. More generally, it probably should extend to a semiotic approach, using Bertin's (1983) visual variables as a first cut at a classification of graphic approaches. In her dissertation, Jois Child (1984) applied such an approach to road maps in general and to one road map in particular, identifying general aspects of the way symbol type, colour, shape, etc. represented different dimensions of meaning. For example, on a Washington State road map, salmon and trout hatcheries were both represented by small fish-shaped symbols, with red for salmon and green for trout. Child noted that most industry-related symbols were red, and that most recreation-related symbols were coded in green. The colour coding of the hatchery symbols reinforces the idea that, in Washington, salmon are chiefly associated with the fishing industry, and trout with sport fishing. Child's work should form a basis for further work on this topic.

Digital objects

The third major class would be the digital objects: points, pixels, line segments, images, chains (polylines), polygons, strips, quadtree quadrants, whole quadtrees, etc. Note that these programming 'objects' include not only digital approximations to geographical entities, but also more abstract data elements such as pixels and quadtrees. As noted in the introduction, this chapter will not discuss this class further. However, any rule base for map generalization would, of course, have many rules that would apply to this domain.

Rules for map generalization

Expert systems are composed of an inference engine, a rule base, and a database. These principles have been discussed elsewhere, in both a cartographic context (Buttenfield and Mark 1991) and in general (see e.g. Brownston *et al.* 1985), and will not be repeated here. However, one aspect not reviewed by Buttenfield and Mark, namely certainty factors, is briefly reviewed here.

Certainty factors and importance factors

Principles of map design require that features should be included or excluded from the map based on their importance with respect to the purpose of the map. Here it is suggested that importance can be modelled as a certainty factor. Certainty factors are numbers between 0 and 1, used in other expert systems (see Winston 1984: 187–92), to model and propagate uncertainty through inference rules. In principle, the cartographic importance should be interpreted as follows: a feature with an importance of 0 would never be shown, whereas one with an importance of 1 would always be shown, given that there is sufficient map space to include it. Relative values of the importance factor would be used to arbitrate between features as they compete for space on the map.

Rules for the generalization of a bay

First, a hypothetical set of rules for defining the importance of a bay along a shoreline is presented. Note that these rules are expressed in terms of entities; computation of relevant parameters (such as bay area) may involve access to digital objects such as chains, but that is not relevant to the rule-maker, who expresses the rule in the geographical domain. A set of computational procedures is required for identifying a bay as a seaward-facing concavity of the shoreline; the procedure would also delimit the bay, dividing it from adjacent features. Work that may eventually lead to the automation of such feature identification procedures is discussed by Buttenfield (1991, Ch. 9 this volume). It is also necessary to determine whether or not the feature is 'important' with respect to the current map purpose. One might propose that sufficiently large bays are inherently important for almost any kind of map. Size might be defined in several ways, for example on the basis of area or on the basis of 'depth':

$$
\begin{array}{llll}
\text{IF} & (\text{Area}(\text{bay}_x) & > & \text{min\_bay\_area}) \\
\text{THEN} & \text{bay}_{x,\text{importance}} & := & 1
\end{array} \qquad [6.1]
$$

$$
\begin{array}{llll}
\text{IF} & (\text{Depth}(\text{bay}_x) & > & \text{min\_bay\_depth}) \\
\text{THEN} & \text{bay}_{x,\text{importance}} & := & 1
\end{array}
$$

Procedures would be needed to compute such objective geometric properties as area, and depth, presumably from coordinates or pixels at the data-structure level.

San Francisco Bay, however, gains its importance not only from its geometric size, but also from its relation to a major city. In a sense the bay 'inherits' importance from a city through a spatial relation, which in this case is represented by the English preposition 'on'. This could be represented with the following rule:

$$
\begin{array}{llll}
\text{IF} & (\text{On}(\text{bay}_x, \text{city}_y) & == & \text{True}) \\
\text{AND} & (\text{Importance}(\text{city}_y & == & \text{High}) \\
\text{THEN} & \text{bay}_{x,\text{importance}} & := & \text{Importance}(\text{city}_y)
\end{array} \qquad [6.2]
$$

Importance can be 'inherited' from other kinds of geographical entities in similar ways:

IF	(Into(bay$_x$, river$_y$)	==	True)	
AND	(Importance(river$_y$)	==	High)	[6.3]
THEN	bay$_{x,importance}$	:=	Importance(river$_y$)	

Examples of cultural features from the US Geological Survey

Interesting examples of rules for selection and symbolization during the compilation of topographic maps can be found in an unpublished document entitled 'Instructions for stereocompilation of map manuscripts scribed at 1 : 24 000', attributed to the USGS's Topographic Division, Rocky Mountain area (USGS 1964). The document was certainly not intended to form a 'rule base' for computer mapping, but rather represents 'common-sense' guidelines, placed in an 'objective' form in an attempt to achieve consistency across large numbers of cartographers working in a production environment. Similar documents from other mapping agencies have been examined by McMaster (1991, Ch. 2 this volume) in the case of the US DMA, and by Nickerson (1991, Ch. 3 this volume) for the Canadian National Topographic Series (Canadian Centre for Mapping).

In this section and the next, many of these rules or guidelines are quoted, and some of them are expressed as explicit rules. Also, commentary will focus on aspects of these guidelines which are difficult to state as simple production rules. Finally, the variability of geometric thresholds will be tabulated and discussed.

Roads and trails

The USGS document contains a number of rules regarding width properties of roads and trails. These will not be discussed here, as they seem more to be a matter of symbol choice than of generalization. Other rules relate to length-based decisions for inclusion.

> Private roads, access roads, and driveways less than 500 feet (152.4 m) in length will not be shown unless of landmark value in areas of sparse culture. (USGS 1964: 35)

> All streets in populated places will be shown regardless of length. (USGS 1964: 36)

The parts of a rule base that would implement these guidelines might look something like this:

IF	(Type(road$_x$)	==	'private_road)
AND	(Length(road$_x$)	<	500 feet (152.4 m))
THEN	road$_{x,importance}$	:=0	

IF	(Type(road$_x$)	==	access_road)
AND	(Length(road$_x$)	<	500 feet (152.4 m))

(rule continued on next page)

THEN	road$_{x,\text{importance}}$	:=	0	
				[6.4]
IF	(Type(road$_x$)	==	driveway)	
AND	(Length(road$_x$)	<	500 feet (152.4 m))	
THEN	road$_{x,\text{importance}}$	:=	0	
IF	(Type(road$_x$)	==	street)	
AND	(In(area$_y$, road$_x$)	==	True)	
AND	(Attribute(area$_y$)	==	populated_place)	
THEN	road$_{x,\text{importance}}$	:=	1	

Recall that, in principle, a feature with an importance of 0 would never be shown, whereas one with an importance of 1 would always be shown.

Some other rules for roads are:

Traffic circles and cloverleafs will be drawn to scale except that the minimum dimensions for road width are to be held. (USGS 1964: 36)

Trails will be shown on map manuscripts after consideration of their importance as a means of communication and their value to the field completion engineers. (USGS 1964: 37)

Railroads

Unlike the road examples outlined earlier, this statement about railroad yards would be very difficult to represent as an 'if–then–else' rule:

Within the [railroad] yard, main-line through tracks are shown correctly placed, but other tracks are symbolized, preserving as much as possible the distinctive pattern presented by the yard. (USGS 1964: 44)

Expressions such as 'symbolized', or 'preserving . . . the distinctive pattern' would be very difficult to specify. A set of several (perhaps a dozen?) rules would be needed to recognize a railroad yard in the first place, to characterize the 'distinctive pattern' within it, and then to 'preserve' that pattern while reducing the complexity, or number, of tracks shown. This is a case which requires a typification operator (McMaster and Shea 1988; Shea and McMaster 1989). However, to be effective, such an operator would have to solve the problems noted above.

Spur tracks, sidings, switches, and storage tracks are mapped accurately as to length, but may be adjusted positionwise [*sic*] if the map scale and adjacent detail require it. (USGS 1964: 44)

This rule calls for an application of a lateral displacement procedure, as described by Nickerson and Freeman (1986) and by Mackaness and Fisher (1987), but otherwise largely ignored in the computer mapping literature.

In congested or urban areas, minor industrial sidings may be omitted, but in sparsely settled country, sidings or short sections of double track are usually landmarks, often have names, and should always be shown. (USGS 1964: 44)

Turntables will be plotted to scale except that they will be scribed no smaller than 120 feet (36.6 m) for 1 : 24 000 scale. (USGS 1964: 46)

These last two principles are similar to several of the rules for roads.

Bridges and viaducts

These also have minimum length thresholds, relaxed by 'importance' principles:

Bridges (and viaducts) on roads and railroads are shown by symbol if they are 300 feet (91.4 m) long or longer. They are symbolized regardless of length if they are drawbridges, if they are important historically, if they are the only crossings in areas of minimum culture, if they are conspicuous in shape or design, or if they are outstanding for any other reason. (USGS 1964: 46)

This could be represented by about six rules, two of which are given below:

IF	$(\text{Length}(bridge_x)$	$>=$	300 feet (91.4 m))
THEN	$bridge_{x,\text{importance}}$	$:=$	1

[6.5]

IF	$(\text{Type}(bridge_x)$	$==$	drawbridge)
THEN	$bridge_{x,\text{importance}}$	$:=$	1
(etc.)			

Urban areas

Urban areas have a minimum size threshold, which is required before a pink tint is used, and minor buildings are suppressed:

The [urban] area must be no less than 0.75 of a square mile (1.9 km^2). Isolated islands adjacent to large urban areas may carry the urban area tint when they are as small as 0.25 of a square mile (0.6 km^2). (USGS 1964: 54)

The rule fails to specify what 'adjacent' means; here, it cannot mean mathematically defined contiguity, since then the features in question would not be 'isolated islands'. Evidently, there is some intermediate range of distances (greater than zero, yet less than some maximum range) for areas to qualify as an 'isolated island adjacent to' other entities.

An initial practical problem in the implementation of this rule would be that of identifying 'urban areas', if the basic data consisted of parcels and buildings. The guidelines go on to say that:

The Technical Planning Unit will make a tentative outline of all areas under consideration for urban tint. (USGS 1964: 54)

Evidently, the identification of such urban areas is considered too difficult, or indeterminate, to allow individual topographers to delineate them. An expert system to perform the tasks of a topographic technician might require that urban areas be identified interactively by a skilled operator. And, since there probably are far more topographic technicians than there are members in a technical planning unit, replication of the tasks performed by the lower-level workers is probably of highest priority during automation. This is consistent with the claim of Herb Schorr, an IBM Vice-President, that encoding the expertise of low-level employees is the most important area of application for expert systems technology in business computing (Schorr 1988).

Open pits

These rules seem straightforward, once an open pit has been identified:

Excavations smaller than 400 feet (121.9 m) on their major dimensions are shown by symbol only. (USGS 1964: 57)

Medium size pits are those larger than 400 feet (121.9 m) but do not exceed 1200 feet (365.8 m) on their major dimension. Contouring is the preferred treatment of these features when they clearly delimit the features. (USGS 1964: 57)

Large pits are those that exceed 1200 feet (365.8 m) on their major dimension. These are always contoured and outlined on the brown plate. (USGS 1964: 58)

The greatest diameter is easily computed, and then must be compared to the thresholds 400 and 1200 feet (121.9 and 365.8 m):

IF	(Length(open_pit$_x$)	<	400 feet (121.9 m))
THEN	open_pit$_{x,size}$	:=	'small'

ELSE IF	(Length(open_pit$_x$)	<	1200 feet (121.9 m)) [6.6]
THEN	open_pit$_{x,size}$	:=	'medium'
ELSE	open_pit$_{x,size}$	:=	'large'

Then, a separate set of rules could be used to determine the appropriate graphic representation:

IF	(open_pit$_{x,size}$	==	'small')
THEN	Plot(open_pit$_x$,symbol)		

IF	(open_pit$_{x,size}$	==	'medium') [6.7]

(rule continued on next page)

OR   (open_pit$_{x,size}$ ==   'large')
THEN   Plot(open_pit$_x$,contour)

Fences

All fence lines discernible on the aerial photograph should be plotted except: (1) Those lines closely paralleling other linear features, such as roads, railroads, canals, and transmission lines. (2) Those lines within or adjacent to built-up areas. (3) Those lines less than 1000 feet (304.8 m) apart are usually not compiled unless they are necessary to show a landmark fence or fence pattern. (USGS 1964: 63)

Implementation of the first exception would require the determination of the spatial relations to other features (e.g. 'closely paralleling other linear features'). And, it is not clear whether 'closely' is intended to be a 'hedge-word' that relaxes the parallelism requirement, or whether it indicates a critical distance, which is not specified. The second antecedent condition is easily handled by polygon enclosure; the third exception requires external knowledge of local importance.

Wells, tanks, and cemeteries

These cultural features have rules based on size:

The (61.0 m) minimum practical spacing between well centers is about 200 feet (61.0 m) for mapping for 1 : 24 000 scale. (USGS 1964: 64)

In congested areas, tanks may be shown with a forty-foot (12.2 m) minimum circle size. (USGS 1964: 66)

Cemeteries 50 feet square or smaller are shown with a solid outline and labeled. Those larger than 50 feet (15.2 m) are outlined to scale with the dashed line and labeled. It is not necessary to label those larger than 200 by 200 feet (61.0 m), but a cross should be placed within the outline. (USGS 1964: 71)

As in the case of open pits, cemeteries are range-graded into three size classes, based on the 50-foot (15.2 m) and 200-foot (61.0 m) thresholds, and then represented or symbolized accordingly:

IF   (Length(cemetery$_x$)   <   50 feet (15.2 m))   [6.8]
THEN   cemetery$_{x,size}$:=   'small'

ELSE IF (Length(cemetery$_x$)   <   200 feet (61.0 m)
THEN   cemetery$_{x,siz_e}$:=   'medium'

ELSE   cemetery$_{x,size}$:=   'large'

```
IF          (cemetery_{x,size}         ==     'small')
THEN        Plot(cemetery_x, solid_outline)
AND         Label(cemetery_x)

IF          (cemetery_{x,size}         ==     'medium')
THEN        Plot(cemetery_x, outline_to_scale)
AND         Label(cemetery_x)

IF          (cemetery_{x,size}         ==     'large')
THEN        Plot(cemetery_x, outline_to_scale)
AND         Plot(cemetery_x, symbol)
```

Examples of naturally occurring features from the US Geological Survey

Foreshore–offshore features

The section of foreshore and offshore features differs from many others in that it contains an interesting set of rules for resolving conflicts between information from two distinct sources. USGS maps 'include certain hydrographic information obtained from recommended nautical charts of the U.S. Coast & Geodetic Survey' (USGS 1964: 75). It is possible that the transferred information might conflict with interpretations that topographers have made from the aerial photography.

> Where conflict is found between the two, as regards position or shape of features, the photogrammetric delineation should be adhered to. In the matter of classification, however, information contained on the charts will be accepted unless the photographs clearly show such classification to be in error. (USGS 1964: 76)

> Rules for handling conflicting multiple representations of the same feature might be handled as data are added to a database, but would otherwise have to be handled by a cartographic expert system.

Islands

Unlike almost all other map features, islands are shown regardless of size. This may be because arbitrarily small islands are still potential hazards to navigation.

> All islands of whatever size will be shown and their shorelines carefully drawn. If an island is so small that its shoreline must be drawn solid, it must be definitely identified by a scribed notation consisting of an arrow and the abbreviation of the world 'island' to 'Is.'. (USGS 1964: 78)

Drainage

The problem of representing drainage is critical for hydrologic or geometric studies based on maps.

> For complete drainage portrayal, all drainage 2000 feet (610 m) long or longer should be compiled. . . . The only channels not plotted are the smaller ones in finely dissected terrain where scale limitations make some selection necessary. In these areas the more prominent stream channels should be plotted and the shorter less important ones omitted. Drainage should not be drawn nearer than 500 feet (152 m) to the divide. (USGS 1964: 80)

The implications for earth and environmental sciences, and the relations to channels actually present in the field, have been reviewed and discussed by Mark (1983). We may question whether the potential loss of geomorphological information, in order to achieve cartographic design objectives, is justified.

Lakes and ponds

This would involve rules for selection and exaggeration (McMaster and Shea 1988; Shea and McMaster 1989).

> Minimum size for lakes and ponds is 65 feet (19.8 m) for 1 : 24 000 scale compilation on the least dimension. Enlarge those ponds that are smaller to that dimension. (USGS 1964: 83)

Canals, ditches

It seems unlikely that cross-sectional area information would be available in most geocartographic databases. Thus, the implementation of this rule would require the entry of data from other sources into the supporting database.

> In general, ditches and canals less than 8 feet (2.4 m) square in cross-section are omitted for 7½-minute maps. (USGS 1964: 89)

Sand areas

This is another case where exaggeration is used to maintain cartographic clarity:

> Sand areas above mean high water level, 0.25 square miles (0.6 km^2) or larger for 1 : 24 000 compilation scale, are to be outlined and labeled

'Sand' or 'Sand Dunes' as appropriate. Sand areas or dunes smaller than the above-mentioned dimensions are to be shown when they are landmark features. (USGS 1964: 96)

Sand washes over 80 feet (24.4 m) wide for 1 : 24 000 compilation scale will be shown double line. . . . However, in areas where sand washes become landmark features, those that are as narrow as 60 feet (18.3 m) may be compiled by exaggerating their width to 80 feet (24.4 m) on the planimetric plate. (USGS 1964: 96)

Tops and saddles

This is a particularly interesting case. The utility, and even the validity, of such an approach differs, depending on the purpose of the generalization.

. . . the most troublesome situation is encountered where the relief along the top of the ridge falls within the range of a single contour interval. . . . The use of a series of spot elevations would show the variations along the ridge, or the exercise of 'topographic license' by deliberately drawing the contours at other than their specified elevations, could serve to indicate the series of tops and saddles at some sacrifice of accuracy. (USGS 1964: 103)

The distinction proposed by Brassel and Weibel (1987), between cartographic generalization and statistical generalization, is critical. If the purpose is statistical generalization, to produce a simplified digital elevation model (DEM) for analytical applications, then the above rules should not be applied. And, if a DEM is constructed by digitizing a map produced using this rule, then the DEM contains biases. Similarly, display of topography using techniques such as block diagrams or analytical hill-shading should also not be generalized using such a rule. But, if the topography is to be displayed from a DEM using contours, then such a rule may be highly desirable.

Woodland

Whatever the map scale, woodland areas larger than the equivalent of approximately 200 feet by 200 feet shall be shown. Conversely, clearings within woodland areas, sometimes important map features, shall be shown if they are larger than approximately 200 feet by 200 feet (61.0 m). This size limit means an area of about 40 000 square feet or one acre (0.4 ha), regardless of its shape. (USGS 1964: 121)

Submarginal coverage:

Land having scattered growth of trees or brush is difficult to classify. As a general rule, if less than 20 percent of the ground is covered, it is called clear. If coverage appears to be 35 percent or above, then a classification of scrub or timber should be made. The treatment of the areas that have a coverage between 20 and 35 percent will be the most difficult. (USGS 1964: 122–3)

Geometric constraints and thresholds

The reader may have noticed that many of the above rules and guidelines contain specific size thresholds, and furthermore, that the specific thresholds vary widely. Table 6.1 lists the quantitative thresholds contained in the rules quoted above. In fact, the data are not inconsistent with the idea that every threshold may have been picked independently; the only values that recur are 200 feet (61.0 m), found three times, and 500 feet (152.4 m), found twice, and these themselves are round-numbered small-integer fractions of an inch at 1 : 24 000 scale (200 feet (61.0 m) equals 0.10 inch (2.54 mm) on the map, and 500 feet (132.4 m) is 0.25 inch (6.35 mm).

There are many obvious questions here: Are these thresholds optimal for their phenomena? Can any of them be justified quantitatively? If so, which? Could some categories share thresholds, thus leading to a smaller set of threshold parameters? Even if the rules are most appropriate for USGS

Table 6.1   Size thresholds mentioned in the USGS rules

Feature	Attribute	'World' size (feet)	Map size (inches)	Map size (cm)
Cemetery	Side length	50	0.025	0.064
Lakes and ponds	'Size'	65	0.033	0.084
Sand washes	Width	80	0.04	0.10
Railroad turntables	(Diameter?)	120	0.06	0.15
Well centres	Spacing	200	0.10	0.25
Cemetery (for cross within)	Side length	200	0.10	0.25
Bridges	Length	300	0.15	0.38
Excavations (non-symbol)	Length	400	0.20	0.51
Some kinds of roads	Length	500	0.25	0.64
Source-to-divide (for streams)	Distance	500	0.25	0.64
Fences (to show individually)	Spacing	1000	0.50	1.27
Excavations (for brown outline)	Length	1200	0.60	1.52
Drainage	Length	2000	1.00	2.54

Feature	Attribute	$\sqrt{}$'World' size (feet)	$\sqrt{}$Map size (inches)	$\sqrt{}$Map size (cm)
Woods, clearings	Area	200	0.10	0.25
Sand or sand dunes	Area	2640	1.32	3.35
'Islands' of urban areas	Area	2640	1.32	3.35
Urban areas (to show with tint)	Area	4570	2.28	5.79

1 : 24 000 scale products, do the map-scale versions of the thresholds apply effectively to other USGS map scale series? Or to topographic maps for other countries? Or to road maps? Or to thematic maps? Even if all of the above rules are reasonable in principle, and could be implemented, there are many open practical questions in their precise implementations.

It is interesting to compare these thresholds with those presented by Nickerson (1991, Ch. 3 this volume) for Canadian topographic maps at a scale of 1 : 250 000. Many thresholds really apply as distances on the map. When expressed as sizes at published map scale, thresholds for some features (lakes, streams, excavations, woods) are very similar, whereas for others (e.g. built-up or urban areas) the thresholds are quite different. In fact, smaller urban or built-up areas shown as such on Canadian 1 : 250 000 scale maps are shown in that way on USGS 1 : 24 000 scale maps (USGS 1964). The minimum real-world sizes are 2000 by 2000 feet (37 ha) and 0.75 square mile (194 ha), respectively.

Discussion and prospects

After a brief review of some of the object definitions and approaches needed to implement phenomenon-based procedures for map generalization, this chapter concentrated on a discussion of a set of instructions for workers compiling and drafting 1 : 24 000 scale topographic maps of the USGS. These instructions contain many statements that can be cast informally as rules, and that human operators could probably apply consistently with a minimum of interpretation or operator variability. Then, many of these rules are discussed in terms of their entry into a rule-based (production) system for map design. Some of the rules seem straightforward, once appropriate forms of representation for the objects and the rules have been adopted. Others contain a much higher level of indeterminacy. Many rules would require spatial relations, patterns, or external measures of 'importance', as well as geometric descriptors. Also the need to establish, inherit, and propagate importance factors through the rule base adds further complication to the research agenda.

The rules include a great many geometrical thresholds based on distances and areas. The variety of these thresholds suggests that they may be arbitrary, and perhaps far from optimal. If the rules were implemented with a test data set, experiments could be performed that would alter the threshold in one rule, while leaving other things constant; results would then be evaluated by expert cartographic supervisors. If possible, common thresholds could be established for groups of rules, and could be represented with a small number of global parameters, rather than as specific numbers 'hard-wired' into individual rules.

Implementation of these rules within a rule-based expert system would be far from trivial, and yet appears to be achievable. It is also necessary for the empirical evaluation of geometric thresholds and of rule interactions. Of course, even if such a system for the automatic compilation of relatively large-scale topographic maps could be achieved, the extensions to other scales, to road maps, or to maps in general, would represent an ongoing series of challenging problems.

Acknowledgements

This chapter represents a contribution to Research Initiatives 3, Multiple Representations, and 8, 'Expert system for cartographic design', of the National Center for Geographic Information and Analysis, supported by National Science Foundation grant SES-88-10917; support by NSF is gratefully acknowledged. I also wish to thank the organizers (Barbara P. Buttenfield and Robert B. McMaster) and the other participants at the symposium in Syracuse, New York for their valuable comments, and thank McMaster and Buttenfield for their constructive help in their role as editors.

Part III

Formulation of rules

7

Constraints on rule formation

M. Kate Beard

Introduction

Automation of a process typically requires the formulation and development of some well-defined and unambiguous rules, but cartographic generalization does not lend itself well to this endeavour. Several factors impede the formulation of rules for generalization. First, generalization has traditionally been practised as an individual artistic skill and therefore incorporates subjective components which do not readily decompose into logical rules. Formalizing the subjective elements is difficult and tends to sacrifice unique and creative aspects of map making.

A second impediment arises in tailoring generalization for a specific map purpose. Rules effective for the generalization of one map type may not be effective for another. Rules for the generalization of a soil map, for example, may not be transferable to the generalization of nautical charts. Development of a common rule base therefore potentially loses sensitivity to requirements of a particular purpose or application.

A third difficulty arises in responding to variation in the spatial and non-spatial characteristics of the geography being represented. Rules should be responsive to local context, considering the spatial and attribute relationships of neighbourhoods of objects and not simply objects in isolation. Spatial and attribute relations among objects, however, can be very diverse, with each variation requiring a slightly different generalization decision or rule. Figure 7.1 and a rule developed by Nickerson and Freeman (1986: 551) for their rule-base generalization system illustrates the potential problems which can occur. Their rule states:

> If a lake or pond would have an area less than the minimum polygon area, do not include it unless there are five or more other small lakes within a radius of 10 mm, in which case replace these small lakes with one small lake slightly larger than the minimum polygon area.

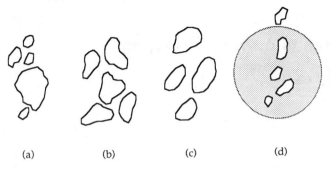

(a)   (b)   (c)   (d)

Fig. 7.1 Hypothetical spatial distribution of lakes

Figure 7.1 shows a distribution of five lakes exhibiting different configurations of size, shape, and spacing and presumably other descriptive characteristics such as depth or turbidity. In case (a), the stated rule will cause the lakes to be replaced by one lake exceeding the minimum area threshold. Figure 7.1(b), despite the different size and spacing configuration, generates the same result: the five lakes are replaced by one lake. Figure 7.1(c) shows a similar size and space configuration to (b) but absence of a fifth lake will cause no lakes to be shown. Figure 7.1(d) illustrates potential problems in defining the appropriate neighbourhood for consideration. In this case a fifth lake falls just outside the 10 mm radius specified by the rule and thus causes none of the lakes to be shown.

Cartographers have the ability to respond to variation in spatial and attribute characteristics and modify rules accordingly. The computer has no comparable ability to modify a rule and so must treat each one literally. The subtle variations in spatial and attribute relationships which cartographers respond to cannot be easily accommodated without generating a sizeable number of additional rules.

Sensitivity to spatial and attribute variation, to map purpose, and map scale suggests the need for flexible or modifiable rules. Standard rule-based systems use the syntax:

If <predicate : THEN <consequence>

with the predicate being a condition or combination of propostions about the database and the consequent, a collection of actions (Shea 1991, Ch. 1 this volume; Armstrong 1991, Ch. 5 this volume). This structure inhibits flexibility since it always binds a condition to a specific action. It is therefore not an optimal foundation for automating generalization. This chapter proposes an alternate approach designed to accommodate flexibility. In the proposed approach constraints substitute for rules. A constraint is a condition similar to the predicate in a production rule. The distinction is that a constraint is not bound to a particular action. The overall rule is that all constraints must be satisfied or resolved, but any number of actions can be applied to resolve them.

Constraints derive from generalization controls such as map purpose, scale, data quality, and graphic limits as given by Robinson *et al.* (1984) and more recently by Brassel and Weibel (1988). Constraints which can be objectively defined from these controls are specified a priori. Remaining constraints can be specified interactively by users and varied to reflect different objectives or purpose. An example of an a priori constraint is the minimum size for legibility. An object violating this constraint requires some action, but several actions are possible. The offending object may be removed, enlarged, or merged with nearby objects to create a larger object. The choice of action depends on the spatial neighbourhood, map purpose, importance of the object, or other variables of significance to the user. This approach is not fully automated but the degree of automation can be increased by increasing the number of constraints specified a priori. The cost is some loss of flexibility.

Figure 7.2 illustrates a conceptual framework for the approach similar to the Brassel and Weibel (1988) concept. A database is assumed to exist. As in the Brassel and Weibel model, generalization controls form the basis for defining a set of constraints. Specified constraints are then matched against conditions of the database. A materalized view (Hudson 1990) of the database is created in which all constraints are satisfied by some set of actions. Actions must either preserve conditions if they already match constraints or correct conditions if constraints have been violated. The end result is a generalized view of the database which satisfies all specified constraints. The remainder of the chapter describes components of the approach in greater detail. The first section focuses on types and formulations of constraints, the second on the actions or generalization operators which must preserve or satisfy constraints. The third section discusses the connections between contraints and actions.

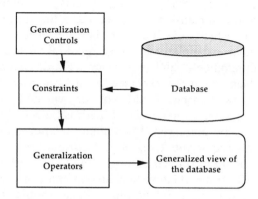

Fig. 7.2 Overview of the conceptual framework for constraint-based approach

Types of constraints

Constraints must be identified which are pertinent to generalization, and one must also identify connections between constraints, and methods for their specification. The constraints are organized into four types: graphic, structural, application, and procedural. These have some overlap with the knowledge areas suggested by Armstrong (1991) in Chapter 5 of this volume.

Graphic constraints

Graphic constraints derive from display configurations and the bounds of human visual acuity: the graphic limits described by Robinson *et al.* (1984). These may be dictated by the physical limitations of display devices such as the size and resolution of display monitors, the resolution and media formats of hard-copy devices, and pen widths in the case of pen plotters. Examples of these constraints include the minimum size and width of objects which can be displayed or the minimum spacing between objects that avoids symbol collision or overlap. All objects selected for a generalized representation are tested against these constraints and any objects violating constraints are tagged as requiring some action. Generalization operations or other actions are then used to correct violations.

Structural constraints

A typical objective of generalization is to capture the essential character of some phenomena and remove unnecessary spatial and attribute detail. The objective of structural constraints is to capture and maintain essential characteristics by capturing and maintaining spatial and attribute relationships among objects. Structural constraints may be expressed in terms of maximizing, minimizing, or maintaining certain relationships. In the spatial domain, distance relationships among objects are fundamental (Beard 1988). Both interior dimensions of objects and the spacing between them are relationships which may need to be preserved or in other ways constrained. Other examples of spatial relations include direction, connectedness, containment, and adjacency. Roads have a connectivity relationship which could be specified as a structural constraint. As a constraint, this relationship could not be violated by generalization operations. For example, when a set of roads is 'selected' their connectivity relationships must be preserved. Maintaining relationships in the attribute domain is equally important. Examples of attribute relationships would be order relations between ordinal data or equivalency relationships between nominal data. The order relationships among cities in terms of population size is an example of a relationship which should be preserved by generalization operators.

Application constraints

Application constraints are conditions specific to a map purpose. Examples of these constraints include location of the geographical area to be displayed, size of the area to be displayed, size and scale of the display format, information content to be included, and types of symbols. Because most application constraints cannot be anticipated prior to generalization they should be interactively specified by users with knowledge of an application's requirements.

Procedural constraints

Procedural constraints control the order and interaction of actions or operations and the order in which constraints are satisfied. Procedural constraints could be specified a priori, but their conditions could change depending on the state of a generalization session. An example of a procedural constraint would be that the 'select' operator precede all other operators. At the beginning of a session only the 'select' operator would be active and others would not be operative until a set of objects had been selected. Procedural constraints would be satisfied by executing the required preceding or supporting actions. Figure 7.3 provides a refinement of the conceptual framework illustrating the interaction and connection of the four types of constraints.

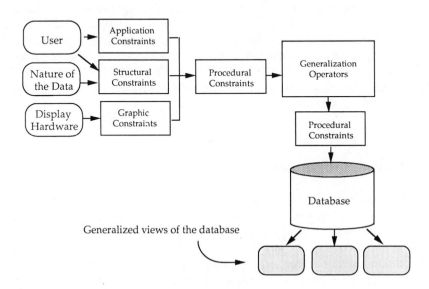

Fig. 7.3 Refining the constraint-based framework

Interaction of constraints

Conflicts can occur between the structural, graphic, and application constraints. A conflict between application and graphic constraints could occur when objects selected by users for a particular purpose could not be legibly displayed by a particular display device. Priorities on satisfaction of constraints must exist to resolve such conflicts when they occur. In the above example the user has two options: to respecify the application constraints or choose a different display configuration (change the graphic constraints). Most users will not have the luxury of several display devices to choose from, so their options for varying graphic constraints can be limited. More often they will be required to modify their application constraints. In general, the graphic constraints will be fixed and application constraints will be variable.

Structural constraints in most cases will override application constraints, but must satisfy graphic constraints. If an application constraint specifies inclusion of a set of roads, the selected set could be expanded by the need to satisfy the connectivity (structural) constraint. Similarly, if an application constraint specifies cities for display, a structural constraint between a city and an island ('contains' relationship) will require inclusion of the island. Structural constraints need not be permanent. If a road and a coastline are essentially parallel, maintenance of the parallel relationship could be specified as a structural constraint or not depending on a user's objective. If specified as a constraint, however, a generalization action such as 'simplify' would be required to preserve the relationship. Structural constraints in the attribute domain are more likely to vary with application constraints. An application constraint could require changes in the equivalency of nominal attribute classes (equivalency in this case being a structural constraint between classes). As an example, an application may require generalization of a soil map to display productivity. The application, in this case, requires a new structural constraint specifying equivalency between soil mapping units sharing the same productivity value. This new equivalency constraint could be satisfied by a 'classify' operator.

Specification of constraints

Graphic constraints can be specified by methods similar to device drivers in which the characteristics of a device are made known to the software. Depending on which display device is selected for output, the appropriate set of graphic constraints would become active. The structural constraints may be handled in different ways. Some structural constraints may be built into the data structure of the database. Topological representations, for example, build in adjacency and connectivity relationships. Structural relations which are application specific can be specified interactively.

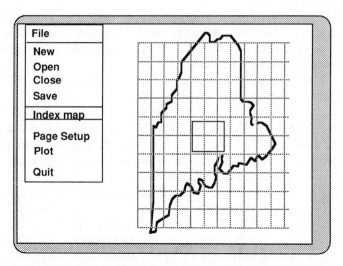

Fig. 7.4 Example interface in which the user is presented with an index map from which to select a geographical area to be generalized

Application constraints are intended to be defined by a specific map purpose and a desired display format. Unless the application is highly standardized, specification of application constraints will require user interaction through a supportive and intelligent user interface. As described above, application constraints can include specification of a certain area, information on a certain theme, objects of certain size, or symbols of a certain type and colour. An appropriately designed user interface could allow users to make these specifications quickly and intuitively. Figure 7.4 shows an example interface in which the user can easily specify a

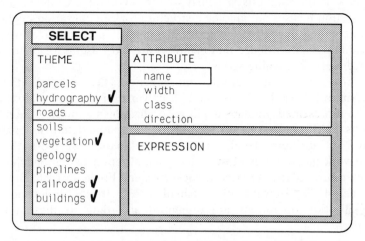

Fig. 7.5 Example user interface for selecting desired information. Information could be selected by theme, class, instance, or any attribute of these

geographical area by pointing and clicking or dragging. The area a user selects becomes a constraint which is matched against the database. The constraint is satisfied by copying from the database the information associated with the selected areas.

Figure 7.5 shows another sample interface panel in which the user is presented with the information available in the database for their selected geographical area. In this display users can again point and click to select desired information. This selection becomes a constraint specifying the information content to be included in the generalized view. This set would be subjected to graphic and structural constraints as well and satisfaction of these constraints could possibly change the original selection.

The application and to some extent structural constraints express desired conditions for a generalized view. The graphic, procedural, and several structural constraints express required conditions. It is the application constraints or desired conditions which in many cases create conflicts or violations of the other, particularly graphic, constraints. When violations of constraints occur they should be flagged and made known to the user. Actions would then be taken to correct the violations. The next section discusses types of generalization actions.

Specification of actions

Actions are operations applied to a database to correct, satisfy, or preserve conditions specified by constraints. In the context of this approach, the function of actions must be clearly defined to anticipate or predict how they will interact with constraints. For example, when constraints are violated we need to know what actions can be taken to correct them. Similarly we need to know that conditions specified for preservation are indeed preserved. This section focuses on building a structure for generalization operators that organizes them according to specific actions they perform. This is analogous to building the knowledge that hammers are most useful for pounding nails, screwdrivers for turning screws, etc.

A first step should consider those actions that generalization should perform at a basic level. Assuming a universe of geographical objects that can be represented, humans might generalize this by limiting the number of objects which are considered, simplifying the spatial detail, simplifying non-spatial or attribute detail, or some combination of these three. This framework incorporates most of the generalization operations described in the literature (McMaster 1989a; Nickerson and Freeman 1986; Brassel 1985; Lichtner 1979; Steward 1974; Rhind 1973). In the following sections, generalization operations are organized according to these three basic functions.

Operations to reduce the number of objects

Operations in this category generalize by reducing the number of objects. In the pure case no change should occur in the spatial and attribute domains, that is, the operator performs one specific function. Examples of such operators are the omission/selection operations identified by many carto-graphers (Keates 1989; Brassel 1985; Robinson *et al.* 1984; Steward 1974).

Specifying operations to reduce the number of objects requires defining the objects of interest. Cartographic generalization typically operates on graphic symbols such that a reduction of objects involves a reduction in the number of symbols. Early digital representations mimicked the graphic symbols on maps so digital representations of points and lines were removed in much the same way as line symbolds were erased from maps. No logical connections existed between symbols (or their digital representations), so the removal of one symbol had no effect on any others. These representations, of which cartographic spaghetti and CAD graphic primi-tives are examples, are referred to here as graphic objects. The distinguishing characteristic of this representation is that generalization operators can be applied to objects independently.

In the digital context, a reduction in the number of objects does not imply their removal from the database. Assuming a universe of objects in the database, all of which can be independently selected for display, a reduction in the number of objects can be accomplished by simply selecting a desired set from the database. The term 'select' will be used here as it is widely used in the cartographic literature in this context.

A selected set can be further reduced by omitting objects from the set. 'Omit' is therefore another operator which reduces the number of objects in a generalized view. A 'select' operation should precede 'omit' (a procedural constraint).

A reduction in the number of graphic objects can also be accomplished by replacing several objects with one new object. For example, independent house symbols may be replaced by one urban area symbol as shown in Fig. 7.6. Several cartographic generalization operations fall under this category

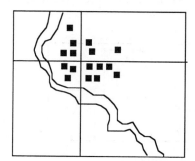 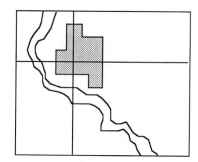

Fig. 7.6   Example of the 'group' operator. In this case the operator groups several house objects into one urban object

(McMaster 1991, Ch. 2 this volume). To avoid confusion with other terminology this operation will be referred to as 'group'. This type of operation does not just reduce the number of objects but typically causes a change in the attribute domain as well.

In contrast to a graphic representation, reducing the number of objects in a topological representation is controlled by structural constraints. Primitive objects in a topological representation include nodes, chains, or polygons (NCDCDS 1988) each of which is logically linked to form regions of a partition at a higher level. These objects may be removed (or not displayed) only if their logical connections and rules for maintaining topological consistency are preserved (White 1984). For example a node must be bounded by chains. Therefore, if the number of nodes is reduced, the number of chains must also be reduced to maintain consistency. Similarly, areas or polygons are bounded by chains. A reduction in the number of chains will therefore result in a reduction in the number of polygons. The higher-level regions also cannot be removed without violating planarity constraints. In other words, a region itself cannot be removed as it would create a hole in the partition. A region can only be transformed into some new region. Operators to reduce the number of objects in a topological representation, being bound by structural constraints, perform differently. It may therefore be necessary to define operators specific to this case such as 'link', defining an operator which reduces the number of nodes and chains. Another example would be 'aggregate', defining an operator which reduces the numbers of nodes, chains, and polygons (Fig. 7.7).

These operations are not limited to the action of reducing the number of objects. Both operators create changes in the spatial domain and the 'aggregate' operator requires changes in the attribute domain.

Operations for simplifying the spatial domain

Operations in this category generalize the spatial definition of objects. Simplification is the term from the literature commonly associated with this operation (Keates 1989; Robinson *et al.* 1984). In the pure case attribute

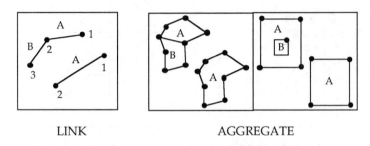

LINK   AGGREGATE

Fig. 7.7 Illustration of the 'link' and 'aggregate' operators

descriptions and the number of objects are not affected. As with the previous operations, the structure of a digital representation can affect performance of these operations. In the graphic context, the cartographer simplifies spatial information by removing small crenulations and curves from line symbols. Modifying line symbols also causes modification of the shape and dimensions of areas if the line symbols serve as representations of areal boundaries. In digital representations, simplification of spatial detail is often associated with removal of points – a reduction in the number of objects (Douglas and Peucker 1973; McMaster 1987). However, if we assume lines and polygons as primary objects, simplification of the spatial detail can be achieved without changing the number of objects. Lines and areal boundaries can also be simplified by applying smoothing filters (Zoraster, Davis and Hugus 1984). The term 'simplify' is used to refer to operations which modify only the metric information.

The spatial domain can also be simplified by modifying the topology. Changes in the topological structure are subject to the same conditions as apply to reductions in the number of objects. That is, the topological primitives cannot be modified independently of each other. In this case the spatial generalization operations can be the same as those for reducing the number of objects, namely 'link' and 'aggregate'.

Spatial information can also be abstracted by reducing the topological dimension of an object representation. In this case polygons or 2-cells may be reduced to 1-cells or 0-cells. Examples of this would be the conversion of a two-dimensional representation of a river or road to a one-dimensional representation. Similarly a two-dimensional representation of a city could be reduced to a zero-dimensional representation. This operation will be referred to as 'collapse'. This type of operator has been described by Nickerson and Freeman (1986), Beard (1987), and McMaster (1989a).

Operations for simplifying of the attribute domain

Operations in this category involve simplification or abstraction of the attribute domain. Again in the pure case such operations would modify only the attribute domain but leave the number of objects and spatial definitions intact. A primary example of this operation from the cartographic generalization literature is classification (Robinson *et al*. 1984). Data abstraction methods from the computer science literature share some of the same terminology and a similar set of operations for abstracting non-spatial information. These operations include classification, generalization, aggregation, and association (Smith and Smith 1977; Brodie 1984; Peckham and Maryanski 1988). Classification in this context is an abstraction process which creates a generic class from individual entities or instances (i.e. mud lake, green lake, and clear lake are transformed to the class lake). Generalization is defined as the combination of two or more classes into a superclass based on shared characteristics (i.e. the classes lake, stream, and

Fig. 7.8 Illustration of the 'classify' operation on nominal and interval data. Under these operations more detailed classes or refined intervals are mapped to less detailed classes or coarser intervals

pond are transformed into the superclass waterbodies). Aggregation in this context is defined as the relationship between several low-level objects which form a higher-level object (i.e. city is an aggregation of component objects such as streets, buildings, etc.).

Assuming all attributes must be measured by one of the four scales of measurement (nominal, ordinal, interval, or ratio), generalization of attribute data can involve simplifying them along either their original scale of measurement or by transforming between scales. For example, the soil attribute permeability, measured on a ratio scale, can be transformed into the ordinal scale of low, medium, and high permeability resulting in a generalization of that information. Attribute simplification along the same scale will involve a reduction in the number of classes for nominal attributes, a reduction in the number of ranks for ordinal data, or a change in the resolution of the interval in the case of interval or ratio scale data (Fig. 7.8).

These operators can operate strictly in the attribute domain. That is, abstraction of the attribute information need not cause a spatial transformation. If, however, attributes are explicitly associated with spatial objects (a structural constraint exists) changes in the attributes will require changes in the spatial structure. For example, if two classes are merged (residential and commercial become urban) the corresponding spatial objects must be merged. In a topological representation, for instance, a 'classify' operation could automatically trigger the 'aggregate' operation (Fig. 7.9).

Connections between operations and constraints

The next step is to examine constraints and their association with the generalization operations described above.

Fig. 7.9 Example in which the 'bisects' relation is preserved during a COLLAPSE operation

Constraints and operations for the reduction of objects

Operations to reduce the number of objects interact with constraints in several ways. The graphic constraints set bounds on what information can be physically and legibly displayed. Applicable constraints include bounds on symbol width, symbol size, symbol length and symbol spacing. If these constraints are violated, the 'select', 'omit', or 'group' operators can be applied to remove objects which are too small, too narrow, too close, or too short. Structural constraints typically specify conditions which need to be preserved. These conditions influence reduction in number of objects by specifying spatial and attribute dependencies between objects. For example, if the 'contains' relationship between two objects is constrained both would be removed by the 'omit' operation.

The application constraints are variable, but can influence the number of objects by setting bounds on the area to be depicted, the set of objects desired, or the themes which are desired. Application constraints may require modification for compliance with the graphic and structural constraints. For example, if the set of objects selected for display are too small to be legible given a certain display configuration, these objects must be dropped from the application specifications or, alternatively, the constraint on the size of the area to be depicted could be revised.

In topological representations the constraints apply similarly. Specific graphic constraints can include bounds on the size of regions and lengths of segments. Segments not meeting the length criteria will be 'linked'. Regions or polygons not meeting the criteria of minimum size constraints could be 'aggregated' or 'grouped'.

Constraints and operations for simplification of the spatial domain

Graphic, structural, and application constraints control generalization operations on the spatial domain. Graphic constraints could include symbol size, symbol width, the radius of curvature for line symbols and spacing between symbols. The 'simplify' operator could be applied to correct violations of these constraints. The execution of 'simplify' must satisfy conditions of minimum distance separation, minimum radius of curvature, etc. In most cases structural constraints would specify conditions to be preserved during spatial modifications. Examples of such constraints could include preservation of right angles for building corners or the parallelism of two features. The 'bisects' relationship between a road and city could be a structural constraint requiring preservation by the 'collapse' operator.

Application constraints would specify reductions in spatial detail desired by users. To support specification of these constraints, users must be provided a mechanism to describe and define extraneous detail. At a minimum the application constraints specified by a user must meet the graphic and structural constraints. Typically graphic and structural constraints would have priority over application constraints.

Constraints and operations for simplification of the attribute domain

Application constraints may dominate in generalizing the attribute domain as more often the generalization of attributes is specific to an application. Some graphic constraints, however, do apply. These can include the range of symbol types, number of colours, number of grey shades, or number of available patterns. These display variables will constrain the number of attribute classes or interval ranges to those which can be legibly displayed and detected as different. A 4-bit colour display, for example, could limit the number of attribute classes to 16.

Other graphic constraints will not directly affect attribute generalization, but require it indirectly by creating changes in the spatial domain. If, for example, the graphic constraint for minimum size is satisfied by deleting boundaries between regions, then an attribute change is required. For example, if two or more regions have been 'aggregated' their names and/or descriptions must be modified accordingly ('re-classified').

Table 7.1 summarizes the different constraints by the operators. This is not an exhaustive summary, but a sample of possible constraints affecting different operators. The sequence of user-specified actions and constraints should be logged during a session. By logging these, the set of constraints and actions creating a generalized view could be captured and used to re-create the same map at a later time. The advantage of this particular approach is that it allows for subjective input, but assures a replicable result.

Table 7.1 Summary of graphic, structural, and applications constraints by operator

Operators	Graphic constraints	Structural constraints	Application constraints
'Select'	Symbol width, symbol length, symbol spacing	Spatial and attribute relationships	Desired area Desired objects Desired themes
'Simplify'	Symbol width, symbol length, symbol spacing	Spatial relationships	Desired level of spatial abstraction
'Collapse'	Polygon size or width	Spatial relationships	Desired level of spatial abstraction
'Link'	Segment length	Spatial relationships	Desired level of spatial abstraction
'Aggregate'	Polygon size or width	Spatial and attribute relationships	Desired level of spatial and attribute abstraction
'Classify'	Colour range, grey scales, patterns, symbol types	Attribute relationships	Desired level of attribute abstraction
'Group'	Colour range, grey scales, patterns, symbol types, symbol size, symbol spacing	Spatial and attribute relationships	Desired level of spatial and attribute abstraction

Summary

Representing data at different levels of abstraction is consistent with the way in which people view the world. If generalization is only automated to create a small set of predefined products it does not serve us well. To be most useful it should allow users to create any generalized view which serves their needs. Such a goal places rigorous demands on the design of an automated process and difficulties in developing a finite set of rules. The approach outlined in this chapter does not try to anticipate all rules in advance, and tries to avoid the inflexibility of standard production rules. Four types of constraints are proposed which can capture conditions required for a particular map purpose, scale, and limitations of display device configurations. Additional flexibility is achieved by not binding constraints to actions. The type and sequencing of generalization operations are only bound to the requirement that all constraints must be satisfied either by preserving conditions if originally true or correcting conditions which have been violated. Some constraints can be specified a priori such as those associated with graphic devices. Others are specified at the start of a generalization session by user specification of desired conditions. A further possibility exists to tailor the approach to different levels of user sophistication. That is, more constraints could be specified a priori for less sophisticated users.

Acknowledgement

This chapter reports research that is part of NCGIA Initiative 3, Multiple Representations and is supported in part by a grant from the National Science Foundation (SES 88-10917). Support by NSF is gratefully acknowledged.

8

Rule selection for small-scale map generalization

Dianne E. Richardson and Jean-Claude Muller

Introduction

Automated generalization, like manual generalization, revolves around the following activities: simplification, displacement, exaggeration, combination, selection, classification, and symbolization. Whereas those activities are applied both to the graphic elements and the thematic content of maps, their involvement is not equally distributed. For example, simplification, displacement, and exaggeration are tools directed towards the manipulation of the graphic presentation of spatial objects, and their intervention is strongly but not exclusively motivated by legibility requirements. On the other hand, selection and classification affect primarily the thematic content of maps, and their involvement is prompted by the type of information which must be displayed. Hence, the former activities emphasize form, whereas the latter emphasize content. Those classes of activities are strongly connected, however, since selection and classification lead necessarily towards changes in graphic presentation and the needs for graphic transcription may in turn promote further selection and classification.

Previous efforts in automated generalization have concentrated mainly in graphic manipulation and object representation. These efforts were prompted by the need to maintain map legibility in the process of scale reduction. Examples of those efforts are the automated generalization of cartographic lines (McMaster 1987) and roads and settlements (Meyer 1987).

Corresponding criteria to judge the performance of map generalization methods were exclusively geometric. Those methods were limited to isolated objects (such as lines) taken out of context and were mainly applied to large-scale topographic maps. They emphasize the 'how' to generalize rather than the questions of 'when', 'what', and 'why' (Shea and McMaster 1989; Richardson 1988). In so doing, automated generalization was breaking away from the long tradition of manual generalization where geographical analysis and evaluation of map content were constant guiding inputs to the process. Recent research efforts have demonstrated the need for an expanded, more integrated approach to the problem of automated generalization (Buttenfield and DeLotto 1989). New efforts must attempt to resolve the following:

1. The development of automated generalization procedures extending through the entire map production range, including large- and small-scale topographic and thematic mapping;
2. The implementation of phenomenon-oriented procedures wherein the substantive meaning of objects as well as their spatial juxtaposition with other objects are taken into account; and
3. The adoption of comprehensive solutions including algorithmic and rule-based approaches and the simultaneous involvement of vector and raster data structures. This would require the ability to interface between such programming environments as FORTRAN, C, and PROLOG.

In order to achieve these goals, new expertise must be gained in the area of rule-based generalization, a relative newcomer in the kit of tools that may be used for map production. One of those tools is rule-based selection for the production of base maps at small scales, and will be discussed in this chapter. The discussion will be preceded by a brief introduction to rule-based heuristics and procedural methods.

Procedural methods versus rule-based heuristics

Shea (1991) discusses the generic form of knowledge-based systems in Chapter 1 of this volume, and the following discussion will focus briefly on the perspective of the programming languages underlying those systems. Most computer languages are procedural languages used to program algorithms for the execution of numeric operations. FORTRAN, Pascal, and C and assembly languages are examples of procedural languages. Inference engines are composed of algorithms or rules and may be expressed in the form of IF–THEN statements. These are called conditional statements and are commonly based on string or numeric matching. Most conditional statements contain a sequence of rules that must be executed in an order predetermined by the logic of the program.

Rule-based heuristics are also a collection of IF–THEN statements. In contrast to procedural methods, the conditional statements relate to symbolic matching rather than string matching, where symbols represent facts of reality. Symbols are always related to data; in order words, there is no separation between data and program. Non-procedural or data-driven language programs operate by executing the rules discovered in the set of IF–THEN statements which most closely represent the relationship between the encountered facts. PROLOG and LISP are both data-driven languages relevant for symbolic processing. In PROLOG programs, rules may appear in random order. This is one characteristic of intelligence, since a search strategy for a solution may not always follow the same path.

IF LINE (OBJECT, LEGEND, SCALE)   [8.1]
    AND OBJECTS (MIDDLE HIGHWAY)
    AND LEGEND (DOUBLE LINE)
    AND SCALE (1 : 1 000 000)
THEN GENERALIZE (LINE, SINGLE LINE)

Automated solutions to map generalization have mostly been implemented in the form of algorithms using procedural languages. They reflect a quantitative approach to a quantitative problem, such as the reduction of the number of points in a line or the quantization of a continuous grey-level picture. Rule-based solutions, on the other hand, are more adapted for the qualitative aspects of map generalization. They can emulate the thinking of a cartographer who is confronted by a series of decisions in the combination of objects, the elimination of irrelevant details, or the symbolization of particular features. Such decisions may be formalized to avoid inconsistencies, non-uniformity, and subjectivity. This is assuming, of course, that knowledge of the generalization process can be formalized into a chain of reasoning paths, each leading to a particular decision or priorities under which generalization may proceed.

Procedural and rule-based solutions are not mutually exclusive. A rule may call for another rule which in turn calls for a procedural routine. Conversely, a procedure may lead to a question that must be resolved by applying a rule-based strategy.

The most difficult issue to be elucidated in the construction of a rule-based system is the acquisition and representation of knowledge. Two questions must be addressed. First, what are the elements of knowledge required in order to decide on a particular generalization decision or choice? Second, how should this knowledge be represented in logical form? Various methods are available for representing and organizing knowledge, including logical representation schemes such as first-order predicate calculus, network representation schemes such as semantic networks or conceptual graphs, and structured representation schemes such as script, frames, and objects. This representation is a necessary step towards the construction of a knowledge base and its implementation in a program. Logical representation schemes, for example, may suggest PROLOG programming. Generally a knowledge engineer is not available to translate and formalize the knowledge of a cartographer. Hence, the expert cartographer is usually confronted with the task of formalizing and programming his or her own knowledge, optimally in collaboration with a computer scientist.

Rule-based approach for selection

One of the key elements of generalization in conventional mapping is selection. Selection necessarily involves its inverse, elimination. Selection and elimination of map elements are parts of a series of transformations (Tobler 1966) which lead to an abstract representation or modelling of reality. According to Grunreich (1985), this modelling takes place at two levels. First, modelling occurs during data capture (in digital or analogue forms) through sampling procedures. Data capture involves a primary stage of object generalization and results in the digital landscape model (DLM). Second, modelling occurs with digital (internal) or symbolic (graphic and external) representation of data. Brassel and Weibel (1987) refer to this as cartographic generalization, which results in a digital cartographic model

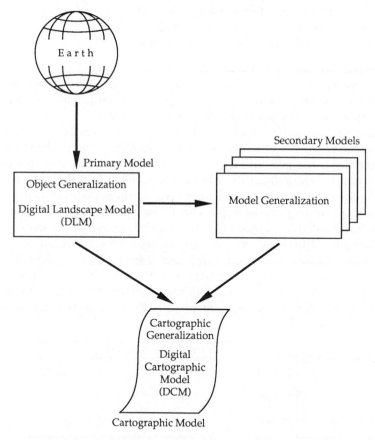

Fig. 8.1 Models derived from the primary digital landscape model (DLM) are special-purpose secondary models of reality. They are free of cartographic representational information. Both primary and secondary models may be used to create a digital cartographic model (DCM) through the process of cartographic generalization (after Grunreich 1985)

(DCM). The DLM and DCM categories are well established in the German cartographic literature and reinforce the idea that generalization is not exclusively prompted by graphic representation (Fig. 8.1).

Selection may be considered as a preprocessing stage to map design and compilation. But selection may also intervene at later stages. Noteworthy is the sequence of generalization processes used in the computer-assisted generalization of topographic maps moving from 1 : 5000 to 1 : 25 000 scale (Lichtner 1979). The sequence runs as follows:

1. Selection of contours and other linear features which are not to be represented at the destination scale;
2. Smoothing of contours and other linear features, for example, axes of highways and railroads, or river channels or shorelines;
3. Displacement and repositioning;
4. Selection and enlargement of buildings;
5. Simplification of building outlines;
6. Combination of individual building outlines; and
7. Renewed simplification of building outlines.

The example demonstrates that selection intervenes before and during the graphic manipulation stages. This points out one aspect of the complexity of the generalization process where the individual transformations (here they are simplification and selection) may occur repeatedly.

A comprehensive rule-based approach to generalization would have to include the possibility of a rule calling itself recursively when the results of a transformation are not satisfactory or have later been adversely affected by other types of operations. Normally, the objective of a rule-based selection system is to provide recommendations as to which map element must be deleted and which must be kept. This in itself is an ambitious task if one considers all individual objects (a stream, a bridge) or parts of an object (the barn unit of a farm) which may be represented on a map. Selection rules for large-scale (1 : 1000 to 1 : 5000) topographic mapping revolve more around geometric than substantive characteristics. For example, object size is more likely to be affected than its significance relative to other map objects. Selection rules for smaller-scale maps (1 : 200 000 and smaller) emphasize

RESOLUTION	REPRESENTATION TYPES		
	SINGLE LINE	DOUBLE LINE	AREAL MARKS
Graphic	0.05 mm	0.35 mm	0.65 mm
Ground	25 meters	175 meters	325 meters

Fig. 8.2 Graphic resolution is determined by the minimum size of visually distinct lines or marks on the map. Ground resolution is determined both by graphic resolution and scale. Here, corresponding ground resolutions were calculated for a scale of 1 : 500 000

substantive characteristics, since at smaller scales, map resolution does not so often permit pictorial or proportional representation of individual objects (Fig. 8.2). Selection rules between these scale ranges involve a combination of both geometry and substance.

Experiment in rule-based selection

The remainder of this chapter reports an empirical study of selection rules for small-scale base maps. The experimental design included three parts, including development of criteria for rule-based selection, development of the rule-based system for map compilation, and assessment of the results.

Development of criteria

Selection criteria were developed for the production of Canadian base maps at very small scales, ranging through 1 : 2 million, 1 : 7.5 million, 1 : 12.5 million, and 1 : 30 million. Those base maps are to provide the background information for 44 different thematic realms, including physical themes (geology, climatology, forestry, etc.), human themes (population, languages, religion, etc.), economic themes (agriculture, communication, employment, etc.), and historical themes (settlement, ethnography, culture). Figure 8.3 shows these themes. They correspond to the themes displayed in the National Atlas of Canada. Selection criteria are to be applied to 14 different

DISCIPLINE	SUBJECT REALMS		
Physical Geography	Geophysics	Geology	Geomorphology
	Climatology	Hydrology	Pedology
	Phytogeography	Zoogeography	Ecology
	Environment	Agriculture	Forestry
	Fisheries	Energy	Regions
Human Geography	Settlement	Political	Population
	Ethnography	Languages	Migration
	Vital Statistics	Health	Culture
	Religion	Education	Administration
Economic Geography	Agriculture	Forestry	Fisheries
	Mining	Energy	Manufacturing
	Construction	Transport	Communication
	Finance	Commerce	Tourism
	Employment	Income	Leisure
	Urban	Regions	Economic
Historical Geography	Exploration	Settlement	Political
	Population	Ethnography	Migration
	Culture	International	

Fig. 8.3 Thematic realms of the National Atlas of Canada (Richardson 1988: 54)

Formulation of rules

Base Map Objects for Thematic Mapping	
1. City	8. Rivers
2. Town	9. Lakes
3. Village	10. Islands
4. Unincorporated Place	11. International Boundary
5. Non-unincorporated Place	12. Provincial Boundary
6. Indian Reserve	13. Census Division Boundary
7. Military Reserve	14. Glaciers

Fig. 8.4 Selected classes of map objects for the base maps of the National Atlas of Canada (Richardson 1988: 40)

classes of objects that normally support thematic mapping, by providing a locational and structural framework (Fig. 8.4).

Two types of evaluation were performed in order to establish a selection index for each class of objects. The first involved criteria based on requirements of base map objects for the 4 different scales and the 44 subject realms. Requirements were established by conducting two types of surveys. In the first survey, staff at the Canada Centre for Mapping (CCM) were interviewed to determine the desirability of base map objects in relation to thematic contexts. The second survey was conducted by reviewing the presence/absence of base map objects in 110 existing maps. Assuming that the cartographer's selection solutions closely reflect the user's requirement for background information, the second survey provides an objective means of assessing requirements. The two combined surveys provide a rating for each object class relative to 44 different thematic realms for display at 4 successively smaller scales. Each class of object for each subject realm was rated for the four map scales according to the following categories:

1. *Essential.* The object is needed to link or support thematic components. For example, 'river networks' are essential for mapping drainage basins.
2. *Desirable.* Not essential, but it helps to give orientation or description to the thematic components. For example, including a selection of the object class 'glaciers' on a map for zoogeography can provide some explanation for different mammal distributions.
3. *Questionable.* It could be used but would not be necessary, such as plotting 'glaciers' on a map showing transportation.
4. *Unnecessary.* Either it is unusable, or would readily be seen as illogical to use, such as plotting census division boundaries on a map of phyto-geography.

The results have been compiled into matrices for each scale. Examples are extracted from those matrices to cross-tabulate the relative importance of each object class for a particular theme (Fig. 8.5) and to cross-tabulate the

| Subject Realm Requirement for Base Map Objects at 1:2,000,000 | | | | | | | | | |
Base Map Object	Geophysics	Geology	Geomorphology	Climatology	Hydrology	Pedology	Phytogeography	Zoogeography	Ecology	Environment
City	⊛	O	⊛	⊛	O	O	O	O	O	⊛
Town	*	*	⊛	⊛	O	O	*	O	O	⊛
Village	–	–	O	O	–	*	*	O	O	O
Unincorporated	–	–	*	*	–	*	*	*	–	O
Non-unincorporated	–	–	–	–	–	–	–	–	–	–
Indian Reserve	–	–	*	O	O	O	*	*	O	O
Military	–	–	*	O	O	*	–	–	*	*
Rivers	⊛	⊛	⊛	⊛	⊛	⊛	⊛	⊛	⊛	⊛
Lakes	⊛	⊛	⊛	⊛	⊛	⊛	⊛	⊛	⊛	⊛
Islands	–	–	–	–	⊛	O	O	*	*	*
International	⊛	⊛	⊛	⊛	⊛	⊛	⊛	⊛	⊛	⊛
Provincial	⊛	⊛	⊛	⊛	⊛	⊛	⊛	⊛	⊛	⊛
Census Division	–	–	–	–	–	–	–	–	*	*
Glaciers	*	O	⊛	O	⊛	O	⊛	O	⊛	⊛

⊛ Essential   O Desirable   * Questionable   – Unnecessary

Fig. 8.5 Requirements for base map objects by subject realm (after Richardson 1988: 45)

| | Climatology | | | | Hydrology | | | |
Base Map Object	1:2M	1:7.5M	1:12.5M	1:30M	1:2M	1:7.5M	1:12.5M	1:30M
City	⊛	⊛	O	*	O	*	*	–
Town	⊛	O	*	–	O	*	*	–
Village	O	*	–	–	–	–	–	–
Unincorporated	*	–	–	–	–	–	–	–
Non-unincorporated	–	–	–	–	–	–	–	–
Indian Reserve	O	*	–	–	*	–	–	–
Military	O	*	–	–	*	–	–	–
Rivers	⊛	⊛	O	O	⊛	⊛	⊛	O
Lakes	⊛	⊛	O	O	⊛	⊛	⊛	O
Islands	–	–	–	–	⊛	O	*	–
International	⊛	⊛	⊛	⊛	⊛	⊛	⊛	⊛
Provincial	⊛	⊛	⊛	O	⊛	⊛	⊛	O
Census Division	–	–	–	–	*	–	–	–
Glaciers	O	*	–	–	⊛	⊛	O	–

⊛ Essential   O Desirable   * Questionable   – Unnecessary

Fig. 8.6 Example of change in requirements for base map objects at various scales (after Richardson 1988: 46)

relative importance for a particular map scale (Fig. 8.6). Hydrography and political divisions are the only object classes considered essential for all tested themes, with the exception of unincorporated and census boundaries, which are often considered unnecessary for thematic support. The significance of objects does reduce with changing scale, that is, with scale reduction one does not discover a non-essential object that suddenly becomes essential. Once again, (international) political boundaries and hydrography maintain the highest level of importance throughout the scale change progression.

The second type of criteria were based on functionality and used to determine why and for what purpose an object appears on a map at a given scale. These criteria were applied independent of a particular theme (such as climate or hydrography) in an effort to make the selection process more objective. Functionality of the 14 classes of base map objects were rated according to 5 types of activities generally employed by users in reading a map: orientation, location, enumeration, measurement, and description (Fig. 8.7). Rating categories were identical to those already used for the requirement evaluation. It is interesting to note that all objects maintain a high level of importance for description, with less consistent ratings of signficance given for orientation and location functions.

Base Map Object	Functionality – 1:2,000,000				
	Orientation	Location	Enumeration	Measurement	Description
City	⊛	⊛	⊛	⊛	⊛
Town	⊛	⊛	⊛	⊛	⊛
Village	O	⊛	⊛	⊛	⊛
Unincorporated	O	O	*	⊛	⊛
Non-unincorporated	*	*	–	*	–
Indian Reserve	O	⊛	O	⊛	⊛
Military	*	*	*	O	⊛
Rivers	O	⊛	⊛	⊛	⊛
Lakes	*	⊛	⊛	⊛	⊛
Islands	O	*	*	*	⊛
International	⊛	⊛	⊛	⊛	⊛
Provincial	⊛	⊛	⊛	⊛	⊛
Census Division	O	⊛	⊛	O	⊛
Glaciers	O	⊛	*	*	O

⊛ Essential   O Desirable   * Questionable   – Unnecessary

Fig. 8.7 Functionality of base map object classes at 1 : 2 000 000 (after Richardson 1988: 51)

Development of the rule-based system

Requirement and functionality criteria may be used to define selection rules. The rules are provided by a necessity factor which determines the intensity of generalization by a selection operator. High necessity factor values indicate that a high proportion of objects in a class should be selected, and a small proportion should be eliminated. Necessity factors were calculated from a combination of requirement and functionality. Qualitative assessments of object requirements were first converted into a quantitative scale so that the ratings of essential, desirable, questionable, and unnecessary were given the values 100, 75, 25, and 0 per cent respectively. Qualitative assessments for the five types of functionalities were converted to quantitative ratings such that an object rated as essential at a particular scale was given values of 80, 90, 90, 90, and 100 per cent respectively.

Necessity factors were computed for each class of objects ($i = 1–14$) and each subject realm ($k = 1–44$) according to the formula

$$(NF)_{ik} = \tfrac{1}{2}(R_{ik} + F_i) \tag{8.2}$$

where R_{ik} is the requirement rating for each base map object class ($i = 1–14$) and for each subject realm ($k = 1–44$), and

$$F_i = \tfrac{1}{5}\Sigma f_n \tag{8.3}$$

is the map object functionality rating for base map object class ($i = 1–14$), where f_n ($n = 1–5$) are the five components (orientation, location, enumeration, measurement, and description). Results of these computations can be seen in Fig. 8.8 showing object requirements, functionality, and the necessity factors for a climatology map at scale 1 : 7.5 million. For instance, a necessity factor of 81 per cent given for object type 'town' means that 81 per cent of towns should be retained for display while eliminating the other 19 per cent.

As expected, the necessity factor decreases with a decrease in scale. It establishes a threshold for the selection of each object class in relation to the subject being mapped at any of the derived scales of representation. The necessity factor provides a general guideline for how many objects should appear on a map given a particular subject and scale. It does not, however, specify which objects within an object class must be selected. To compensate for this limitation, a set of rules were developed for each object class to determine which objects should be selected according to the necessity factor value. For example, in the case of rivers on the climatology map, a necessity factor of 87 per cent indicates that 13 per cent of the rivers must be eliminated. This criterion can be met by dropping streams with a low stream order. When a choice is required to drop one or the other of a stream order with the same value it is made according to length at the specified source scale. This process of selection allows the major drainage networks to be retained and the minor tributaries to be eliminated. Because

Necessity Factor for Mapping Climatology at 1:7,500,000			
Object	Requirements %	Functionality %	Necessity Factor %
Cities	100	87	93
Towns	75	87	81
Villages	25	54	39
Unincorporated	0	21	10
Non-unincorporated	0	0	0
Indian Reserves	25	21	22
Military	25	5	15
International Boundary	100	90	100
Provincial Boundary	100	90	100
Census Division	0	50	0
Rivers	100	74	87
Lakes	100	54	77
Islands	0	22	11
Glaciers	25	21	23

Fig. 8.8 Calculation of the necessity factor for base map objects on climatology maps at 1 : 7 500 000

the rules are consistently applied, regional variations in hydrography are preserved (Richardson 1989).

The system was implemented using data captured from 1 : 2 million base maps for the provinces of Alberta and Saskatchewan, Canada. Base map objects were digitized and stored in a relational database using the Masmap/Mimer system of Finmap, Finland. The Masmap system structure consists of a graphics database and attribute database. Mimer, the relational database management system, consists of several modules including a query language used to develop rules. A total of 2915 objects, relating to the 14 classes of map objects, were recognized. Attributes such as area, length, stream order, drainage discharge, etc. were included in the system and stored in relational tables. Attribute information provided the criterion for which objects to select given the necessity factor, what to select according to the requirements of the subject being mapped, and why something should be selected according to the functionality of the individual object types.

Assessment of the results

Evaluation was conducted at two levels: (1) comparison between the maps produced on the basis of the rule-based selection with already existing manually generalized maps; and (2) comparison of the general guidelines provided by the rule-based selection with Töpfer's Radical Law for generalization (Töpfer and Pillewizer 1966). In the first comparison, it was found that manually generalized maps had generally fewer base map elements than the maps derived using the rule base. The second comparison was only indicative, since Töpfer's Law only considers scale reduction without paying attention to subject realm.

In this context, a rule-based approach built upon a good knowledge base is potentially much more powerful than Töpfer's simple formula that was originally developed for topographic maps. Also, this evaluation suggests that this rule-based approach needs further fine tuning according to the results provided by manually generalized maps. Finally, some of the rule-based selection guidelines, presented in the numerical form of a necessity factor, must allow for exceptions in displaying some object classes. For example, in displaying boundary information either all of a provincial or international boundary should be shown or none at all. Decreases in the number of elements within these object classes would not be logical and therefore exceptions must exist within the necessity factor (Fig. 8.8).

Summary and perspectives

A rule-based system may take different forms. In its conventional form, a rule-based system must include a separation between the knowledge base including objects, facts, and rules, and the inference engine which uses a variety of mechanisms such as unification (or symbol matching), backtracking, and backward chaining in order to prove a hypothesis or achieve a particular goal such as generalization on the basis of object selection. In this sense, the example discussed in this chapter does not conform to the standard definition of a rule-based system. It is more like a database that takes the form of a look-up table where each entry is a rule recommending the level of selection that must be achieved for a particular map component or class of object at a particular scale for a particular theme.

This first step attempt towards a rule-based generalization procedure raises a number of interesting issues, however. First, there is the issue of rule definition. A rule is usually derived from previous knowledge or expertise in a particular domain. Knowledge in this case was the outcome of a survey addressed to experts in the field of map making and map use as well as an analysis of manually generalized maps. This involved a series of processes such as the redaction of a questionnaire to determine the requirement and functionality of map objects as well as a personal assessment of the relevance of those objects on already existing maps. That is, a procedure of enquiry was created that reflects personal views about modelling knowledge. Another knowledge engineer might use another questionnaire or another strategy to determine and compute the necessity factors derived from a combination of requirement and functionality. 'No rules exist for how to determine first the requirement of an element to appear on a map nor the assignment of a percentage for that requirement' (Richardson 1988; 61). The absence of rules to determine which rules or facts must be included in a knowledge base is a recurrent problem in building expert systems. Since the outcome of a rule-based system cannot be of higher quality than the incoming information provided by the database, there is critical need for quality control at the input level.

Selection rules expressed through the global percentage of information which must be retained for a particular class of objects do not provide any guidance about which object or part of object must be kept or deleted. Hence, the system includes additional information that establishes the geographical significance of each object. This information is largely attribute information but should also contain contextual information. The attribute information used characterizes the intrinsic quality of an object, such as size, length, growing rate, etc.; whereas contextual information refers to the connections of an object to other objects, such as distance, communication, and density. A centrally connected town may be judged more important to retain than a peripheral place. A small lake in a highly dense drainage basin area may be given less significance than a lake of the same size in a barren area. A fast-decaying glacier may call for more attention than a stable one. Whereas some of the rules relating to attributes are easy to implement, such as the selection of cities by rank order of population, other rules related to context may be more difficult to formulate.

Moreover, flexibility must be provided to bypass the rule base in case of 'anomalies' that could not be covered in the knowledge base. Only rule-based selection for very small-scale maps was discussed in this chapter. Alternative rule-based experiments for generalization could involve other tools such as classification, combination, and symbolization. There is a need to experiment with the connection between rule-based decisions and geometric transformation executed algorithmically in the range of large-scale maps, as well as to identify the impediments obstructing total automation of map generalization (Muller 1989).

The formulation of rules and their implementation is one of the most difficult challenges in the cartographic research agenda of the 1990s. Whether those efforts will be successful is still uncertain. Robinson, in his second edition of *Elements of Cartography* (1960: 132) already alluded to those difficulties: 'Many cartographers have attempted to analyze the processes of generalization, but so far it has been impossible to set forth a consistent set of rules that will prescribe what should be done in each instance. It seems likely that cartographic generalization will remain an essentially creative process, and that it will escape the modern tendency towards standardization.'

Acknowledgement

This chapter reports an empirical study of selection rules for small-scale base maps, conducted at International Training Centre (ITC) Enschede, The Netherlands (Richardson 1988) as thesis research under the supervision of Ir. C. A. de Bruijer and Professor Muller.

9

A rule for describing line feature geometry

Barbara P. Buttenfield

Introduction

A continuing challenge for automating the cartographic process relates to using data from a digital cartographic database for representation at multiple map scales. The challenge involves feature simplification, and specifically the determination of feature details that must either be retained or omitted for appropriate graphic representation. The digital database is often produced for multiple purposes, including mapping at multiple scales; it is increasingly rare that a base map is digitized for mapping at a single scale. A related problem is that tolerance values selected for simplifying base map information must be modified as feature geometry varies within the digital file to ensure both accuracy and recognizability of graphic details on a generalized map. At present, decisions about where to adjust tolerance values are made manually, and form an expensive bottleneck to map production for government and commercial organizations.

This chapter explores a method for generating base map features at many scales from a single digital file, and presents a rule by which to determine those scales at which line feature geometry might be expected to change in map representation. The research has application to automating map simplification, incorporating numeric guidelines into digital files about what magnitude and variation in geometric detail should be preserved as the digital file is simplified for representation at reduced map scales.

Derivation of rules to guide the mapping process has been of long-standing interest to cartographers, primarily for reasons of consistency and quality control. The National Map Accuracy Standard was established in 1947 (Thompson 1979) to ensure horizontal and vertical control on USGS topographic maps. The Radical Law (Töpfer and Pillewizer 1966) provided numeric guidelines by which to determine how much detail to retain during map compilation and reduction. This is one of the earliest published rules formalized for map reduction and simplification. The inclusion of coefficients in the formula to control for map purpose and for dimensionality of treated features attests to the recognition that the appearance of map features depends upon both feature type and map purpose. But because a map surface is not homogeneous in the amount or type of detail it contains, the rule cannot be applied with mechanical uniformity.

For example, a topographic sheet may contain very dense settlement features within an urban area, with rectangular street patterns composed of (uniformly) rectangular geometry. It might be logical to apply a single rule to simplify the street pattern. Another part of the same sheet may lie beyond the confines of the urban area, and contain few settlement features, but perhaps include a drainage channel, or transportation network, or agricultural areas. Here, there may be very few features displaying rectangular patterns, or even uniformity. Coefficients for the Radical Law that are appropriate for one part of the map are not likely to provide appropriate simplification for every part of the map. In every case, the geometry of the map symbols must reflect the geographical structure of the landscape, and vary accordingly during map simplification.

One might argue that the solution to this problem is to simplify first the point features, then the line features, and so on. It can be shown, however, that the problem will arise even when treating line features (for example) in isolation. In digital form, cartographic lines are bundled as features that are not always tied to geographical or geometric uniformity. For example, the outline of the USA may be stored in a small-scale database as a single entity and incorporate both natural (coastline) and artificial (arcs of latitude) portions. Another example is provided by a digitally stored contour line that may contain very different amounts and types of crenulation and geometry as the terrain it crosses varies in bedrock hardness and composition.

The cartographic challenge is to apply simplification operators (rules) that accommodate the geographical and geometric changes occurring along the extent of the base map file. At one level, rules may involve changing the simplification or smoothing algorithm. At a different level, rules may involve changing tolerance values to preserve various geometric characteristics (e.g. line length). In a similar fashion, automating decisions about where to modify either the algorithm or the tolerance value must be based upon recognition of where the feature details can be seen to change in size or density. This creates problems for simplification of base map details, particularly for naturally occurring linear features.

The purpose of this research is to evaluate automatic methods to describe line feature geometry as it varies with map scale. This requires formalized description (knowledge) of the amount and type of details that occur along the extent of the digital file, and knowledge as to the scale at which the feature representation should change. In Chapter 5, three types of knowledge are discussed, including geometric, procedural, and structural (Armstrong 1991); it is the geometric knowledge which is the focus here. This chapter presents a method by which to determine changes in geometry, and demonstrates its application for several small examples.

The need to accommodate scale dependence in geographical depiction has been argued in previous literature, as geographical line features vary in appearance with changing scale of map representation (Mandelbrot 1986; Buttenfield 1984; Mark and Aronson 1984; Carpenter 1981; Goodchild 1980). It will be shown here that such features vary in their graphical geometry as well. Information identifying the type of geometric change and the specific map scales at which that change becomes visually evident can be utilized during map simplification to choose tolerance values that preserve both realism and accuracy of the feature as it is represented at multiple scales. The information can be collected as a formalized rule that can be stored in a digital coordinate file and used by a knowledge-based generalization system.

The rule presented in this chapter is termed a **structure signature**. It is a method of hierarchic subdivision and geometric measurement of a digital line feature. It provides formalized description of the line feature's geometric characteristics at successively finer levels of resolution, to accommodate the issue of scale dependence. The rule will be applied to demonstrate distinctions in three different geometric characteristics common to line features on maps, and to justify the need to break digital line files into smaller pieces to preserve uniform geometry during map simplification. Determination of changes in geometry may be inferred statistically, although the shape of the probability density function on which inferences are based may be non-standard.

It is important to note at the outset that while the following discussion will be expressed primarily in terms of vector coordinates and a vector-based solution, a similar argument (and solution) might be proposed within a raster environment with negligible modifications in analytical geometry. Many existing cartographic data sets (USGS DLG–E, National Ocean Survey (NOS) World Vector Shorelines, and Census Topologically Integrated Geographic Encoding and Reference (TIGER) files, for example) are currently formatted as vector strings or (as in the case of TIGER data) vector links between topological nodes. Many large GIS packages (e.g. ARC/INFO, System9, TYDAC) store feature data in vector form; thus the vector solution seems relevant. Where feasible in the discussion below, examples and references to raster processing will be incorporated into the discussion.

Preservation of details during line simplification

In map simplification, algorithms are applied to digital files to remove unwanted detail, to select or emphasize particular items, or to clarify by removing visual clutter. Most simplification algorithms incorporate some mechanism to control the amount of detail that is removed; for example, in an 'n^{th} point' algorithm, the n refers to a numeric threshold (a tolerance value) determining that $1/n$ points will be eliminated systematically or randomly (Tobler 1966). Tolerance values can take many forms. They provide the width of corridors within which coordinates are eliminated (Deveau 1985; Douglas and Peucker 1973), or the number of coordinates to be considered for conversion to a straight line segment (Lang 1969). As argued above, tolerance values must be modified where it becomes evident that feature geometry has changed. Van Horn (1985) presents an example of the Virginia coastline, demonstrating problems with application of constant tolerance values. Manual tolerance value modification is accomplished intuitively, by visual inspection and reliance upon geographical familiarity with the map feature. Two cartographers will probably simplify the same line feature in nearly the same way, and vary their manual simplification in similar localities. But it will be rare that the identical coordinate location will be marked for transition in manual simplification.

In computer simplification, different solutions may result from the starting bias of a particular algorithm; this is particularly true of the family of 'corridor' tolerancing routines just discussed (Deveau 1985; Opheim 1982; Reumann and Witkam 1974; Ramer 1972). Inconsistent treatment of the same geographical feature in various digital map products can lead to incompatibility of adjacent map coverages, or of data products generated at multiple scales, and compound problems of inter-agency data transfer. The decision of how to choose simplification algorithms or how to select tolerance thresholds that guide their operation is neither well understood (McMaster 1987) nor readily formalized. Visual appearance and accurate positioning of a feature must be preserved, and appearance may vary across a range of map scales. Then, too, variations in size and frequency of detail will occur along the extent of the file, as geomorphology and terrain vary.

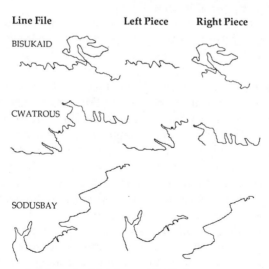

Fig. 9.1 Lines sampled from the McMaster (1983) data set

For example, in Fig. 9.1, the line BISUKAID (1499 points) is a 1 : 62 500 feature whose details differ in amplitude along its extent. The left piece shows smaller crenulations and a more unidirectional trend than the right piece, whose geometry approaches a space-filling curve. The cusps by which the left piece is defined may be eliminated by a simplification algorithm designed to generalize coarser angular details within the right piece. The line CWATROUS (875 points) is a 1 : 62 500 contour extending across differing bedrock material. The left piece exhibits sharp angularity, indicating softer bedrock and more localized downcutting, while the harder bedrock beneath the right half is apparently more resistant to local erosion. Graphically speaking, the details do not differ in size so much as in angularity for this line feature. SODUSBAY (1213 points) represents a coastline (1 : 62 500) and exhibits wave deposition and erosion along its extent. The left piece is characterized by higher frequency or density of detail. (All three line features are drawn from the McMaster 1983 data set.) The research question to be posed in this chapter concerns whether the geometric distinctions (amplitude, angularity, and density) of detail can be identified by formalized descriptions, and whether the formalized descriptions can be implemented as rules for knowledge-based simplification.

Knowledge-based simplification requires that the amount and type of detail in the digital file are defined before the algorithm begins to operate, and that expectations of the amount and type of details that should be retained or eliminated at the reduced scale are also defined. The current cartographic practice is to design algorithms with variable tolerance values that may be modified until the resulting simplification 'looks about right'. For large files containing many features of non-uniform geometry, the current practice is to assign initial tolerance values and then to monitor the progress of the algorithm through a large coordinate file, halting its operation to modify the tolerance threshold.

Returning to the lines in Fig. 9.1, the tolerance value needed to eliminate large amplitude details along the right piece of the BISUKAID file would probably eliminate too much of the smaller amplitude detail along the left piece. Breaking out pieces of the line with different amplitudes of detail and applying different tolerance values to each piece will preserve both the large and small amplitudes during simplification. For a small example such as this, of course, the decision of where to break the line can be accomplished by visual inspection. For large-volume mapping (i.e. coordinate files of 50 000 points or more) this type of manual intervention is inefficient, expensive, and will probably produce inconsistent results during simplification.

Formal description of the geometry contained within a digital file is complicated by the necessity to accommodate scale dependence. Natural features such as coastlines and river channels can be seen to vary in appearance depending upon the scale at which their representation is digitally encoded. One might argue that this is because the geographical feature is continuous, while the digital encoding methods for map representation are discrete. As pointed out by several researchers (Richardson 1961; Steinhaus 1954, 1960; Volkov 1949; Shokalsky 1930), geometric parameters of geographical features obtained by repeated measures using smaller and smaller units of measure do not always converge, and map representations at differing scales must therefore be in some respects unique.

From the context of satellite remote sensing, an example is revised from Buttenfield (1985). If the length of the Puget Sound coastline is measured on a LANDSAT image by counting pixels, its length will tally at some (rough) multiple of 79 m, assuming that image pixels are 79 m on a side. Features of the coastline smaller than 79 m will not be resolved, and thus escape measure. A thematic mapper (TM) image (resolution 30 m pixels) will incorporate some of these features, and the coastline will not be identical to the LANDSAT representation. Its length will be roughly equivalent to the length of the 79 m representation plus the length of all additional features resolved by the TM encoding. A Système Probatoire d'Observation de la Terre (SPOT) image of Puget Sound (10 m resolution) will incorporate still more features. Adding the length of these to the coastline measure will increase the length once again. This process will continue, through resolutions collected by high-altitude and low-altitude raster imaging, down to actual geodetic traverse measuring straight-line distance between selected points. Thus the length and consequently the details of the Puget Sound coast will continue to vary with changes in the scale of their (data capture and) measurement.

Geometric parameters as simple as line length have demonstrated capabilities for crude geomorphic distinctions (Buttenfield 1989), and one would expect that other parameters in addition to line length will serve to refine these distinctions. The formalized description of scale-dependent geometry may be thought of as a set of rules by which the geomorphic logic is preserved during map generalization; the set is comprised of a group of descriptions governing specific geometric parameters. The rules should be encoded into the digital files in a form directly available to the simplification algorithm. An example follows to demonstrate the utility of geometric parameters in a mapping context.

The cartographic importance of parametric description

With decreasing scale, the map representation of a large river narrows from an areal strip to a single line. The parallel indentations of a mapped fjord behave somewhat differently, however. At first, the two sides become more closely spaced, remaining proportional to the scaled width of the fjord. At a certain scale the fjord disappears altogether from the map representation. Rules for depicting fjords and rivers differ in this respect, and cartographers take this into account in deciding at what scale the graphic metamorphosis takes place. Intuitively speaking, two lines can be said to have different

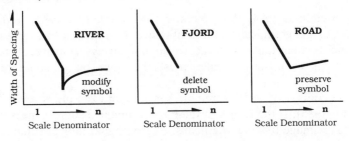

Fig. 9.2 The 'width of spacing' parameter provides a comparison of the relationship between size in graphic and geographical space across a progression from large-scale maps (left side of graphs) to small-scale maps (right side of graphs). The parameter distinguishes scale-dependent behaviour for three types of map features

geometries if they behave differently at different scales. Behaviour in this case refers to the digital values which geometric parameters take on as the representation is depicted at various scales. In differentiating fjord from river in this example, one may differentiate 'width of spacing' between roughly parallel lines bounding the areal strip of river bed or fjorded valley.

It is possible to plot hypothetical spacings for categories of line features at various scales in order to provide a graphic comparison of the relationship between size in graphic and geographical space. Three categories of lines are shown in Fig. 9.2, each of which is symbolized by a double line. The x-axis represents the denominator of the representative fraction, implying larger-scale representations on the left-hand side (e.g. 1 : 50 000) and smaller scale (e.g. 1 : 1 000 000) on the right. The y-axis represents the relationship between line spacing on the map and the actual width of the geographical feature. Negative-sloping lines on the graph indicate that as the scale denominator increases, map spacing decreases in proportion to geographical width. Positive slopes indicate that map spacing is becoming disproportionate to geographical width.

At large scales the river symbol width is proportional to actual width of the geographical feature it represents. With scale reduction, the river width on the map should decrease in linear proportion to the scale change. This kind of generalization is similar to the measurement of river channel width using aerial photography taken at increasing altitudes. Notice that the limits of resolution for the riverbank width will vary for a single channel. For example, the banks of the mouth of the Columbia River will be resolved at much smaller scales than will the headwaters. Thus a single entity in the database may require a variety of cartographic treatments along its extent.

Below the limits of resolution, the river should most appropriately be symbolized by a single line. The relationship between plotted line width and actual geographical width will change suddenly with the single line depiction, as shown by the vertical jog in the plot. The single line may be more narrow than the feature which it represents at the particular scale.

With further reduction (and especially if pen width is kept constant) the relationship between graphic and geographical width will rebound some-what, and then approach some equilibrium at which the cartographer decides to remove the feature from the map. The map scale at which the feature is deleted will be a function of the importance of the river to the particular map purpose.

Line spacing for fjords differs from rivers in cartographic treatment, although at the largest scales the two symbols are treated in similar fashion. With decreasing scale, the width of the symbol is decreased proportionately. At the limits of resolution, however, both lines of the fjord symbol will be removed: by cartographic convention, fjord geometry is not defined by a single line. The convention is reflected by the abrupt termination of the plot.

The third plot stands in juxtaposition to the first two to exemplify cultural features. For fjords and rivers, the double line symbol is irregular at larger scales, reflecting irregularities of terrain. For parallel lines symbolizing a road or highway, the sides of the line symbol will be more regular, to reflect civil engineering standardized specifications of road width, radius of curvature, etc. At larger scales, of course, line spacing remains proportional to its geographical counterpart. At the limits of resolution, the double line representation may be preserved, as with interstate highway symbols on a road map. With further decrease in scale, the symbol is neither deleted nor modified. Line spacing will eventually become 'larger than life' as the map feature becomes disproportionately wider than the geographical road.

An important aspect of all three plots is that each one contains an identifiable jog or elbow. In a topographic mapping situation, one can identify quite specifically the scale at which the jog will occur. Limits of resolution are defined as half the National Map Accuracy Standard, or 0.01 inch (0.3 mm) (1/100 inch) at scale. According to this rule the jog should occur at the scale for which river width, road width, or fjord width reaches 0.01 inch (0.3 mm) on the map. For a 1 : 62 500 quadrangle, fjorded valleys smaller than about 50 feet (15.2 m) across should not be represented. At 1 : 125 000 the threshold is doubled. The resulting effect on rivers, for example, will be symbolization of all channels less than 100 feet (30.5 m) wide by a single line. Muller (1990) has also described evidence of such jogs, which he terms **cusps**, for settlement features generalized on Dutch topographic map series. One goal of formalizing scale-dependent descriptions is to identify specific scales at which cusps are expected to occur, for these are the levels of resolution at which the form of the map feature can be expected to metamorphose during generalization. For scale ranges between the cusps, one can speculate that the map features will change in some proportion to the scale change.

This example describes a relationship between three hypothetical features and their graphic representations at progressive scales in terms of a parameter relating line spacing on the map to geographical width of the feature. The cartographic application demonstrates how parametric rules may be used to predict the structure of map representations for the category called 'river' on topographic maps. The advantage of using parametric rules relates to the flexibility with which slight modifications can be implemented. For example, the cartographic simplification of a freely meandering stream will differ somewhat from the cartographic treatment of a tightly constrained stream channel in preserving the periodicity of meanders. One would expect the rule for rivers will be similar but not equivalent for all classes of rivers, although this remains to be demonstrated.

One is reminded of the techniques applied in a remote sensing training exercise to build a spectral signature for grass or asphalt and then apply it to distinguish land cover types on a satellite image. The spectral signature describes the reflectance pattern for a given type of land cover. This description is then compared with the image to delineate areas having similar patterns of reflectance, and to distinguish between dissimilar land covers. The spectral signature provides rules for expected reflectance across a range of wavelengths. In analogy, the line spacing plots described in this chapter may be thought of as a **structure signature** providing rules for the expected feature geometry represented across a range of map scales (Buttenfield 1984, 1986, 1987).

Other researchers have considered the signature concept in categorizing information at multiple scale (Pike 1988). Varying forms of the concept are referred to by similar names, for example 'fingerprint' (Witkin 1986). The parametric approach and the reduction of continuous data to discrete geometric measurement are common to all, although parameters vary from one implementation to another. In any implementation, a set of parameters (rather than a single measure) will probably be required to capture the full complexity of geographical details. Some parameters will best distinguish certain kinds of features, but not others. It is not the purpose of this research to compare between parametric signature methods, but rather to explore the application of a particular method, the structure signature, to the description of topographic line features. Extensions to line simplification and tolerance value modification will be proposed subsequently.

A parametric rule for scale-based description

The parameters

Five parameters will be incorporated into the structure signature. Each is based on a geometric measurement, for example line length, or zero-crossings (Thapa 1988) of a coordinate string across some sort of anchor line. Measurements are repeated for subdivisions of the line using the Douglas reduction algorithm (Douglas and Peucker 1973). As the line is subdivided, each section is stored in a striptree data structure (Ballard 1981). Measures are made for each strip, and summarized across the tree. This procedure and the parametric measures are refined from Buttenfield (1986, 1989) and Jasinski (1990).

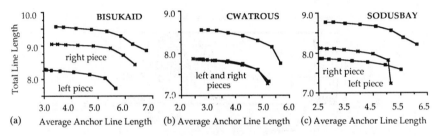

Fig. 9.3 Richardson line length plots provide clearest distinctions between features whose details vary in amplitude, as in the case of BISUKAID. The lack of differentiation for plots of CWATROUS and SODUSBAY indicates that angularity and density of detail are not so clearly distinguished by this parameter

The first parameter (shown in Fig. 9.3) relates the decrease of average strip length at finer levels of resolution to the increase in total line length. This measure was first reported by Richardson (1961), and is the basis for Mandelbrot's (1967) subsequent derivation of fractal dimension. The Richardson plots are used here instead of the fractal *D* value, which has been criticized by a number of cartographic researchers for problems of instability (see Clarke 1990 for a good summary of this work). Buttenfield (1989) modified Richardson's original procedure by relaxing his assumption of linearity. Instead of fitting a linear regression model to the set of measured lengths, she found that merely connecting points in the graph discloses values of average anchor line length at which the rate of increasing total line length changes suddenly. These sudden changes identify cusps, as predicted in the line spacing example above, and were also discovered by Muller (1990).

The resolution (average anchor line length) at which both pieces of the CWATROUS graph (Fig. 9.3(b)) change slope is just under 5.0 units, whereas for the left piece of SODUSBAY (Fig. 9.3(c)) a marked cusp is apparent at a resolution of just over 5.0 units. One should expect that cusps will occur at different resolutions for different line features. This kind of information justifies the avoidance of single tolerance thresholds applied uniformly to simplify all features on a map coverage, for clearly a tolerance value that is sensitive to (that is, eliminates) details smaller than 5.0 units will simplify both the CWATROUS pieces, but have no visible effect upon the SODUSBAY file. A small increase in the tolerance value will generate a very different looking simplification of the map as a whole. Knowledge about the type and amount of details contained in specific features can be used to computational advantage (that is, avoiding the computation time when applying a tolerance value that will have no visible effect) as well as to select a tolerance value that will reduce details more uniformly across the map surface.

While the Richardson line length parameter can provide knowledge about the amount of detail contained in a line feature, other geometric distinctions

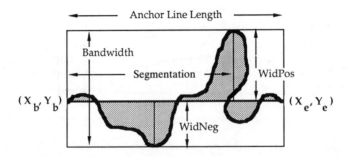

Fig. 9.4 Parameters for a structure signature include the measurements displayed here. The MBR is the rectangle bounding the line segment. Anchor line length is the length of the MBR, and bandwidth is the width. Segmentation is the distance from the beginning coordinate to the location on the anchor line where the maximum deviation occurs. Error variance is the discrete approximation of the shaded area. Concurrence is a count of the number of times the coordinate string crosses the anchor line

are not made clear. One can see in comparing Richardson plots for the pieces of BISUKAID that the line length parameter distinguishes readily between variations in amplitude of detail, although it is not providing clear separations for the changes in angularity found in the CWATROUS file. The varying density of detail evident in the pieces of the SODUSBAY line is evident at finer levels of resolution, where denser (high frequency) details increase overall line length. However, the distinction between line pieces is not as apparent for frequency of detail as for amplitude. Robust distinction between the line files probably requires combining the Richardson line length parameter with other parameters in the structure signature.

In Fig. 9.4, the original coordinate string for a line (or piece of a line) is shown to crenulate around a straight line connecting endpoints of the string. The first computation (anchor line length) measures Euclidean distance between the beginning and ending coordinates of the string.

$$\text{Length} = \sqrt{[(X_b - X_e)^2 + (Y_b - Y_e)^2]} \qquad [9.1]$$

This measure is used to standardize two parameters describing the minimum bounding rectangle (MBR) surrounding the coordinates for each piece of the line as it is subdivided. With further subdivisions, the anchor line length for a line piece will more closely approximate the length of the coordinate string within the MBR. The implication, of course, is that the anchor line provides the most simplified representation of the original coordinate string, which justifies its use to standardize both MBR parameters. The first of these, labelled 'bandwidth', measures the maximum perpendicular deviation of any coordinate in the original string on either side of the anchor line. Deviations are summed to compute the width of the MBR. The bandwidth label is used in the same context as Peucker's (1975) appellation.

$$\text{Bandwidth} = \frac{(\text{WidPos} + \text{WidNeg})}{\text{Length}} \qquad [9.2]$$

In the formula, 'WidPos' and 'WidNeg' represent deviations on either side of the anchor line (the positive side is on the same side of the anchor line as the origin of the coordinate space) (Bartsch 1974: 263) and are illustrated in Fig. 9.4. 'Length' represents anchor line length, as computed in the first equation. The bandwidth parameter additionally provides a measure of the cross-sectional symmetry of the line feature, for equal deviations on both sides of the anchor line indicate a feature with symmetric amplitudes of detail. This measure should therefore provide good distinctions between line pieces exhibiting differing amplitudes of detail, as in the case of BISUKAID.

$$\text{Segmentation} = \frac{\sqrt{[(X_b - x)^2 + (Y_b - y)^2]}}{\text{Length}} \qquad [9.3]$$

The second MBR parameter is called segmentation and is defined as the location along the anchor line where the next subdivision will occur, that is, the location of the coordinate (x, y) which lies at the maximum perpendicular distance from the anchor line. Distance is measured from the beginning coordinate of the anchor line (X_b, Y_b). Segmentation is standardized to the anchor line length to eliminate bias of measurement units. For example, a segmentation value of 0.25 indicates that the maximum deviation occurs ¼ of the way down the anchor line. Interpretation of this parameter may indicate ranges of self-similar geometry. If the segmentation value is preserved across several levels of resolution, that would imply that the line details (at least the maximum deviation) are occurring again and again at the same (relative) location along the line.

Two other parameters describe the path of the coordinate string within the MBR. Error variance is computed in the format of any standardized deviation, as the sum of the squared deviations of distances between coordinates in the original string and the anchor line. McMaster (1986) computes a very similar measure to reflect areal displacement.

$$\text{Error variance} = \sqrt{\left[\frac{\Sigma (\text{Distance})^2 - \dfrac{\Sigma (\text{Distance})^2}{\text{No. of coordinates}}}{\text{No. of coordinates} - 1} \right]} \qquad [9.4]$$

The error variance parameter is illustrated in Fig. 9.4 as the shaded area. Error variance is a discrete approximation of the total discrepancy between the anchor line (the most simplified representation of the line) and the original coordinate string). Error variance is standardized by the number of coordinates in the string, and is reported logarithmically. The formula for distance is expressed in the usual format for directed distance from a straight line in the plane, using the generalized coefficients (A, B, C) for the anchor line equation (Bartsch 1974: 261). Distance of any coordinate (x, y) from the anchor line is given by

$$\text{Distance} = \frac{Ax + By + C}{\text{Length}} \qquad [9.5]$$

The slope of the anchor line will affect the sign of the coefficients A, B, and C, and therefore of the distance value. Positive distance values indicate deviations lying on the same side of the anchor line as the origin of the coordinate space, and negative values indicate deviations on the opposite side of the anchor line. Maximum positive and negative values of the distance computation produce the WidPos and WidNeg values used to measure bandwidth above.

Concurrence, the fifth parameter, is also measured using the distance computation. It is defined as a count of the number of times the distance value changes sign (from positive to negative values) going in sequence along the coordinate string. This indicates the number of times the original string crosses the anchor line, or how closely the anchor line concurs with the original coordinates' path. Concurrence is standardized by the number of coordinates in the string, to give a count of actual crossings as a proportion of the potential number of crossings. This allows the concurrence to vary between 0 and 1, in similar fashion to a correlation coefficient. An arc of a circle will have a value of zero, for example, and the value for a coordinate string in which coordinates alternate back and forth in zigzag fashion about the anchor line will approach 1.0. In Fig. 9.4, the coordinate string crosses the anchor line four times (endpoints are excluded). Thapa (1988) applies a similar measure (referred to as 'zero crossings', to indicate the change in signed values), although he measures the parameter for only a single level of resolution and does not standardize his values.

Incorporating the parameters into the rule

The generation of the structure signature rule involves measuring and summarizing five parameters of a line (Richardson line length, bandwidth, segmentation, error variance, and concurrence) at successively finer levels of resolution. For a digital file, this means measuring the line as a whole, by constructing the anchor line and MBR, then computing (anchor line) length and width of the MBR, concurrence, segmentation and error variance. The line must be subdivided in some consistent fashion and the procedure of measurement repeated, to simulate finer levels of resolution. Subdivision may proceed by means of breaking the line file in two equal-sized portions (in half), or into two randomly sized portions, or by some substantive criteria (for example, choosing as a breakpoint the coordinate marking first-order geodetic control, or the location of a city, major shipping port, or prominent landmark). In this research, the goal is to retain recognizability of the line feature as it is represented at many levels of resolution; thus the line subdivision must be designed to account in some way for preservation of details important for line recognition.

This is accomplished by application of Douglas and Peucker's (1973) line simplification algorithm. Its application in this research is not incidental, for several reasons. First, the line reduction routine has been shown (Kelley 1977) to identify coordinates of maximum angular change, which Attneave (1954) identifies as a major priority for shape recognition. Additionally, the Douglas routine identifies an almost identical set of coordinates as those selected by visual inspection to be critical for recognition of the line in its simplified form (White 1985). The algorithm's selection of these critical points (as Marino 1979 has referred to them) generates simplifications that mimic those generated by manual generalization, and retains details critical for map reader recognition. Finally, the Douglas algorithm can be applied in hierarchic fashion (Douglas and Peucker 1973) to subdivide a line file automatically, while retaining the line's critical shape information overall.

The line files used in this project (BISUKAID, CWATROUS, and SODUSBAY) were subdivided in the following manner. First, points were selected randomly to initialize eight pieces without introducing cartographers' bias. This action might be considered analogous to the identification of certain points in a coordinate file that must be preserved, such as city locations, points where hydrographic channels intersect, map sheet edges, and other constraints beyond the cartographer's control. The five parameters were measured for this (random) subdivision, and stored in a striptree data structure. Each piece was next subdivided using the Douglas algorithm, to produce 16 pieces, which were measured and stored. The 16 were subdivided again using the Douglas-Peucker algorithm to produce 32 pieces. Any given piece of the line within which the maximum deviation fell below 0.0005 inch (0.0127 mm) was not further subdivided, as this is half the resolution at which the lines were originally digitized. When 90 per cent of the line pieces reached the limits of tolerance, the subdivision procedure was terminated.

A final step in generating the structure signatures involves summarizing the measured parameters for each level of resolution. A logical choice summarizes by computing a mean and variance for each measured parameter, as it can be argued that the subdivision process (by any method) is a sampling procedure. Many different line pieces could result from using different subdivision methods. Thus the particular set of line pieces resulting from any particular subdivision is one sample representing the line broken into finer and finer pieces. Use of the first and second moment statistics also provides a good check on the homogeneity of line measurements, that is, the magnitude of the variance may indicate lack of uniform geometry within the line file as a whole. This point will be returned to later in the chapter.

The intention in summarizing parameters for any level of resolution is to incorporate the measurements for all of the pieces, including pieces whose details have been previously resolved. Parameters for line pieces whose details had reached the resolution limits at a previous level of subdivision were incorporated into the summaries for subsequent subdivisions, so that summaries for any single level of resolution incorporated a complete representation of the line file. As discussed earlier in the chapter, it is reasonable to expect non-uniformity in the amount and type of detail occurring on a map. Likewise it is reasonable to find that details within a single feature will also vary, as geometry is not currently used as a criterion for breaking out features in a digital file. Therefore, one can predict that any striptree formed by this procedure will be sparse in some locations. Further discussion of structure signatures may be found in Buttenfield (1984, 1986) and Jasinski (1990). The remainder of the chapter demonstrates a small implementation of the structure signature concept to show how it may be applied to distinguish geometric character.

Calibrating the structure signature rule

In order to formalize a rule for distinguishing between types of geometric variation, such as amplitude, angularity, and frequency of detail, one must calibrate the rule so that parameters measured for features of equivalent geometric character are (nearly) equivalent, and parameters for different geometric character are distinctive. This should hold true for differing features as well as for portions within a digital line file whose geometric character is non-uniform (not homogeneous). If portions of a line feature differ then the structure signatures for those portions should reflect those differences, if the signatures are to be considered useful rules for distinguishing between types of detail to be preserved during simplification.

To demonstrate, nine signatures have been constructed (see Figs 9.5–9.7), three for each of three features taken from the McMaster (1983) data set and shown previously in Fig. 9.1. These features display differences in geometry along their extent. The lines differ in one case in amplitude of detail, in another case in blockiness or angularity, and in a third case the variation in frequency of detail is the distinguishing factor. For the purposes of calibrating the structure signature rule, the line pieces have been arbitrarily broken on the basis of visual inspection. Once it has been shown that the signatures are distinctive, then the question of implementation becomes realistic. Without the 'proof of concept', the rule is difficult to validate.

It is important to realize during the following discussion of the structure signatures that comparisons are limited to pieces of a single line file, not between them. For BISUKAID, comparisons on the basis of amplitude of detail will be emphasized. Emphasis for CWATROUS and SODUSBAY will be based on angularity and frequency of detail, respectively. The digital files were broken into the pieces already displayed in Fig. 9.1, and structure signatures generated for the entire file and for each piece.

Each column of the figure represents a structure signature, including the parameters measuring the MBR and the parameters describing the path of

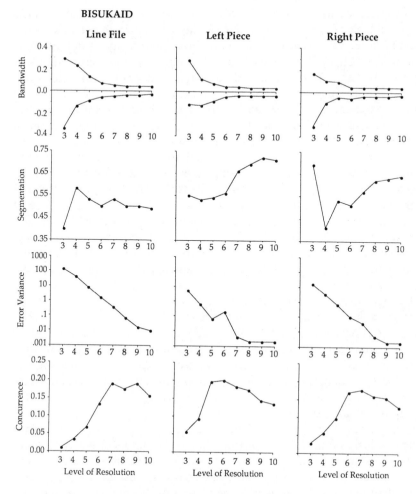

Fig. 9.5 Structure signatures showing differences in amplitude of detail

the coordinate strings within the MBR. Along the x-axis of each plot, tick marks represent levels of resolution at which the parameters were measured. For level 3, the line was subdivided into 8 pieces (2^3), for level 4, into 16 pieces (2^4), and so on, using the Douglas-Peucker routine. Due to the differential details along each line, and the fact that not all of the line files contain 1024 points to begin with, the striptree will become somewhat sparse at finer levels of resolution; however, at least 90 per cent of the nodes are filled at every level, as described above. Signature parameters for the entire line are graphed in the leftmost column, and signatures for the two pieces are displayed in the centre and rightmost columns, respectively.

As expected, some parameters seem to be more effective than others for distinguishing differences in amplitude, blockiness, and frequency of detail. Several parameter graphs display scale-dependent cusps referred to earlier in the chapter, identifying levels of resolution where geometry of the lines changes suddenly. Other line files may display geometric characteristics not considered here. For example, one could study periodicity of details along the coastline of Cape Hatteras and North Carolina, or variations in sinuosity along the British Columbia coastline. This underscores the need for refinement of the signature parameters and for the possible incorporation of additional parameters. The parameters presented here are not intended to be encyclopaedic. Development of additional parameters and refinement of those presented in this chapter form a topic for further research.

It is possible to identify several distinctions in each of the structure signature examples (Figs 9.5–9.7), using some but not all of the signature parameters. In Fig. 9.5, for example, the bandwidth parameter provides good distinction for variations in amplitude of detail. For the right piece of BISUKAID, notice that the maximum negative deviation at level 4 is of a greater magnitude than at level 3. This is unusual, as one would expect smaller amplitude details to be resolved with finer resolution. Probably the larger deviation has to do with the closed shape apparent in the right piece; this geometry is obscured in the signature of the whole file, however.

In the segmentation graphs, one can see that for the line taken as a whole, there is an initial relocation of the maximum deviation, followed by a series of subdivisions occurring roughly about half-way along the anchor line. This implies longitudinal self-similarity, which is not borne out by signatures for the two pieces of the line file, both of whose subdivisions drift along the anchor line for this range of resolutions. It is apparent that non-uniformity of geometric character can be obscured when treating large sections of a cartographic file as a single item.

The cusp in error variance at level 6 for the left piece of BISUKAID indicates an increase in the sum of the squared deviations of the coordinate string from its anchor line. The small amplitude crenulations evident in the left piece appear to be about the same size and of a constant frequency, and level 6 identifies a scale at which they are first becoming resolved. To simplify BISUKAID to this scale, a tolerance value should be chosen that is sensitive to this amplitude of detail. Suitable values for a tolerance threshold can be determined from the y-axis of the bandwidth graph, which displays the average maximum deviation (roughly 0.4–0.6 inch (10–15 mm) on either side of the anchor line) found at this level. Additional information can be computed using the Richardson line length parameters, to determine a resolution (average anchor line length) which will produce a given total line length.

Concurrence has been defined as a ratio between the actual number of anchor line crossings and the number of coordinates in the strip. The ratio can rise even if the number of crossings stays the same, since fewer coordinates will be contained as the subdivision process isolates smaller pieces of the line. It is the level at which sudden slope changes occur that is of interest in this parameter. A cusp from positive to negative slope always indicates a drop in the number of anchor line crossings. At fine resolutions,

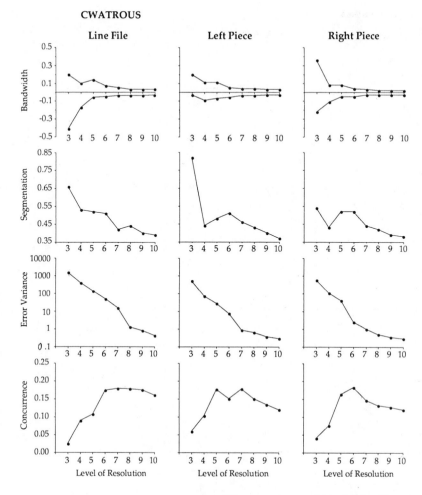

Fig. 9.6 Structure signatures showing differences in angularity of detail

this implies that details in the coordinate string have been largely resolved. For BISUKAID, this occurs at level 7 for the line taken as a whole, but at level 5 for the left piece (smaller amplitude details) and at level 6 for the large amplitude details of the right piece.

For CWATROUS (Fig. 9.6), the line pieces do not differ in amplitude of detail so much as in angularity. Bandwidth for the line taken as a whole and for the right piece display a similar steep drop during early subdivisions, and the smaller magnitude deviations evident in the graph for the left piece are masked by this pattern. Segmentation provides the most distinctive parameters. For the line as a whole, the horizontal slope across levels 4–6 indicates that segmentation is occurring at roughly the same location along the anchor line, implying statistical if not precise self-similarity for the parameter in this range of resolution. Segmentation for the right piece of the

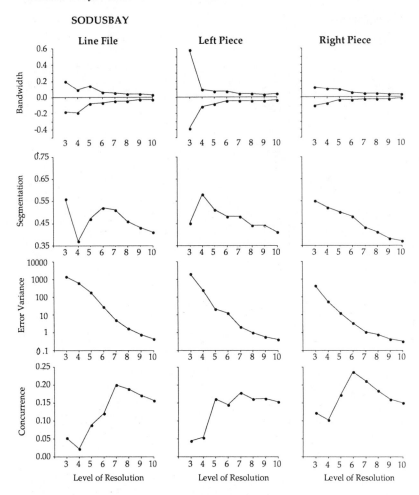

Fig. 9.7 Structure signature showing differences in frequency or density of detail

line displays self-similarity for two of these levels. Map representations generated within this range may appear nearly identical. The pattern of segmentation for the left piece is distinct, and oscillates back and forth along the anchor line. This characteristic is masked completely in the signature for the entire line.

Neither the error variance nor the concurrence graphs display clear differences, and this is logical, as the pieces of CWATROUS differ neither in the overall amount of deviation nor in the frequency of deviation from their anchor lines. Angularity is not so well described by these parameters as by segmentation. Kelley (1977) demonstrated that the Douglas-Peucker algorithm will tend to select coordinate locations of maximum angular change and will also control the selection of the segmentation point. One should expect that the pattern of segmentation will reflect substantial variations in angularity accordingly.

While differences can be seen for several parameters graphed in Fig. 9.7, differences in frequency of detail are expected to be most apparent in graphs of the concurrence parameter. The drop in concurrence at finer resolutions is apparent here as in Fig. 9.5 and 9.6, and again the cusp of change from positive to negative slope occurs at different levels. As a ratio of actual anchor line crossings to potential for crossings, concurrence is a probability measure. The left piece displays a high density of high-frequency details, and the cusp at level 4 where the positive slope suddenly increases indicates a rise in the number of crossings coupled with a drop in the average number of coordinates. For the right piece and line taken as a whole, the cusp at level 4 indicates an initial drop in concurrence; the lower frequency details display a periodic character whose phase may be synchronized with anchor line length at this resolution to form pieces of the line that are like arcs of circles.

Structure signatures for the pieces of the line can be seen to differ from the signatures for the lines taken as a whole. For some parameters, the differences are quite marked. Parameters for Richardson line length, bandwidth, and error variance appear best to distinguish differences in amplitude of detail. Angularity seems to be most clearly distinguished by segmentation, although this may be dependent upon subdivision by the Douglas-Peucker algorithm. If another subdivision procedure were applied, for example simply dividing the line pieces in half, the segmentation parameter would display a very different pattern, obscuring the angularity differences. Frequency of detail seems best distinguished by concurrence. Other parameters may improve the discriminating ability of the structure signature rule, or contribute to the distinction of other types of geometric characteristics.

As proposed in beginning this experiment, a robust rule describing feature geometry should take on unique values to indicate different geometric characteristics. This has been demonstrated. The counter-proposition, that descriptions of similar geometric characteristics should appear similar, has only been alluded to here. A more comprehensive calibration will require demonstration that the same feature collected from different data sources or by different methods (e.g. scanning and vector digitizing) will produce similar structure signatures. This is beyond the scope of the current chapter.

Implications for map generalization

One interesting aspect of the examination of line geometry using structure signatures is that geometric differences evident in signatures generated for pieces of a line are often masked when parameters are averaged for the line as a whole. This argues for generating structure signatures for line features of uniform geometry to ensure the parameters are not biased by heterogeneous amounts and types of detail. This begs the question of how to determine portions of a digital file containing uniform detail. It would be unfortunate if the generation of structure signatures required plotting the entire digital file and marking it manually, for of course this is the very obstacle one is trying to avoid by generating the signatures in the first place. There are at present no formalized rules that may be used to delineate sections of a file containing uniform detail.

Full implementation of such rules will require a certain amount of data exploration. For example, the determination of what is homogeneous geometry will probably differ for cultural features and naturally occurring features. Examples include highways, whose radii of curvature are constrained, and railroads, whose path across terrain is often constrained by gradient. For line features particularly, determination of serial trend may depend as much on map purpose or the map audience, as on geomorphology. Actual determination of geometric distinctions for a particular feature type will probably require some form of empiric evaluation or perceptual testing to preserve consistency in implementation. For now, its utility for automating map simplification must remain speculative.

The structure signature rule presented in this chapter can be implemented now, providing knowledge to automate cartographic line simplification. The structure signature's purpose is to determine scales at which the geometry of a line feature changes. Implementation requires that information be stored with the digital data on the amount and type of details that occur along the extent of the line. That is, line files should be tagged (as one tags coordinate strings with feature codes, for example) with the geometric characteristics that must be preserved within that string at particular map scales. This method can be applied when coordinate files are entered into a database, or as a form of preprocessing. Feature headers could take on form of a look-up table of mean and variance pairs for given parameters, where the look-up table values provide knowledge for selection of tolerance values to preserve specific geometric characteristics. It must be recognized that current simplification algorithms do not look for knowledge of this sort within the data files on which they operate, and thus full implementation of a knowledge-based simplification system will require more than simply adding information to file headers.

The work described in this chapter is intended to demonstrate that rules can be formalized to describe geometry that changes with scale, and to provide information about geometric characteristics that might be retained or eliminated during map simplification tasks. Procedures for generating a structure signature have been described and applied to small cartographic examples to demonstrate its utility for feature descriptions. Indications from this research have been discussed as to how the rule might be refined by modifying parameters. New geometric parameters may be developed to improve overall discriminating ability of the signatures. The structure signatures are presented as an example of a rule by which line feature geometry may be formalized, and applied to break digital lines automatically into pieces that are homogeneous in geometric character. Implementation of this concept has major implications for reduction of production costs and preservation of quality control in using a single digital line file to general maps at multiple scales.

Acknowledgements

This chapter reports research that is part of NCGIA Initiative 3, Multiple Representations and is supported in part by a grant from the National Science Foundation (SES 88-10917). Support by USGS National Mapping Division is also gratefully acknowledged. Comments by reviewers and by the Syracuse Symposium participants have clarified the writing.

10

Amplified intelligence and rule-based systems

Robert Weibel

Introduction

Most of the existing approaches to the automation of map generalization have been algorithmic and rather mechanistic. Furthermore, solutions have only covered certain aspects of map generalization. Research has focused on line simplification, rather than attempting to automate line generalization comprehensively or address the interdependent generalization of related features (e.g. the elements of a topographic map). Consequently, in the past few years quite a lot of authors have argued that automated procedures for map generalization should incorporate more 'intelligence' (Muller 1989; Mark 1989; Brassel and Weibel 1988), and should be more comprehensive or even holistic at the same time. It has also been suggested that such 'processing based on understanding' could best be automated through the application of artificial intelligence (AI) strategies, in particular through expert systems.

Although such arguments have been brought forward for some time now, few actual implementations of expert systems for map generalization have been reported (e.g. Nickerson 1988a; Robinson and Zaltash 1989). This chapter will discuss why approaches based on expert systems – although conceptually promising – have so far had only limited success in this particular domain. There seems to be a fundamental impediment to expert systems approaches, related to a lack of understanding about the processes of map generalization, i.e. a crisis of knowledge engineering for map generalization. A different approach, termed **amplified intelligence**, is proposed: key decisions default explicitly to the user, whose knowledge is amplified by a range of high-level tools for carrying out generalization operations. This approach has two main advantages. As it is more conservative it will lead more easily to the development of operational systems that may be used in productive work. Then too, it allows an analysis of the use of the system functions by human experts. Amplified intelligence should provide a more structured tool for knowledge engineering than conventional techniques. Therefore, amplified intelligence is essentially a transitional approach: knowledge is gradually brought from the human to the system, and eventually this may lead to a full-scale expert system.

The present situation of automated map generalization

One of the most serious limitations to a meaningful use of geographical information systems (GIS) is the current lack of suitable techniques for automated map generalization. Such procedures are needed for two major reasons: to ensure optimal readability of GIS display products, and to enable mapping from, and transformation between, high-resolution multi-purpose spatial databases (to reduce the costs of data capture, increase data consistency, enable cross-database analysis, etc.).

The central importance of map generalization has meant that researchers have addressed this problem from the very beginning of digital cartography. Most of the research, however, has been devoted to rather narrow aspects of the overall problem, such as line simplification. Such an approach was reasonable during the early days of digital cartography. However, as the complexity of cartographic databases is exceeding that of line element databases (e.g. USGS digital line graphs) and as an increasing number of thematic layers are being combined in GIS analysis, more comprehensive solutions are needed. Contemporary research has thus sought to incorporate more knowledge into generalization procedures. Common to many recent publications is the promotion of the application of knowledge-based techniques (in particular expert systems). For further reviews of relevant research in map generalization, see for instance Zoraster, Davis, and Hugus (1984) or Brassel and Weibel (1988). The following discussion is restricted to publications that typify recent tendencies.

A number of theoretical or conceptual papers have addressed aspects of building a knowledge base for automated map generalization. For instance, Mark (1989: 76) has advocated a phenomenal approach to generalization: 'in order to successfully generalize a cartographic line, one must take into account the geometric nature of the real-world phenomenon which that cartographic line represents'. He thus extends the concept of generalization as 'processing based on understanding' (Brassel and Weibel 1988): instead of merely attempting to understand the graphical structure of map elements, the comprehension of their underlying geographical phenomena is demanded. It is interesting to note, however, that such a phenomenal approach to generalization has been promoted and taught in traditional cartography for many years (e.g. Imhof 1965). Apparently this knowledge has been partly lost in contemporary cartography.

A different aspect of generalization has been addressed by Monmonier

(1989c). He proposed a solution for generalization and feature displacement involving the use of interpolated generalization between spatial databases of different scales. This approach is based on the idea that existing geographical databases presumably reflect cartographic skill, since they had been captured from manually generalized maps. This method expands on multi-resolution hierarchical databases proposed by Jones and Abraham (1986). Technically, it resembles key-framing used in computer animation. Due to this familiarity, there is some promise that further animation techniques could be exploited for the definition of key-frames (i.e. map databases at a certain scale) as well as interpolation between frames. The approach, however, seems limited to generalization between map databases of specific scale increments and thematic contents. Thus, it is probably most useful for the production of topographic maps, in particular for map series (e.g. national map series). The production of thematic maps of arbitrary and unprecedented contents (i.e. the graphical results of exploratory GIS applications) can be supported only partly by this technique.

There has also been a tendency towards decomposing generalization into a series of interrelated processes. This strategy has primarily been initiated and applied by European researchers (e.g. Hake 1975; Lichtner 1979; Powitz 1989). It has recently been theoretically elaborated in great detail by Shea and McMaster (1989). They present a detailed framework of 12 generalization operators (i.e. processes) whose execution is initiated depending upon 6 conditions. This divide-and-conquer strategy offers guidelines for theory, communication, and implementation of generalization principles. Shea and McMaster, however, do not try to suggest that generalization could be built just as the sum of individual operators.

Successful generalization depends to a great extent on the harmony between map elements. It is, therefore, important that the interdependencies and sequential effects among the various operators be clearly understood. The study of these problems has not yet received the attention it deserves. One early example has been presented by Lichtner (1979) who postulated a particular sequence of operators for the generalization of large-scale cadastral maps. His scheme was largely based on intuitive judgement and common sense. Although feedbacks between different operators were possible to some limited extent, their order could not be modified. More recent projects have also used predefined operator sequences (e.g. Nickerson 1988a; Weibel 1989a). Monmonier and McMaster (1991) have attempted to develop a more comprehensive model. Their intention was to model sequential effects in line generalization depending on external constraints such as map purpose and scale, and also allow feedbacks among operators. For instance, the authors propose different sequences of operators for the generalization of topographic source data containing boundaries and transportation features depending on the priority order among the two feature classes. Future research will have to back up concepts of this kind by empirical studies in order to obtain more specific guidelines for the sequencing of generalization operators.

The relatively large quantity of theoretical papers addressing aspects of knowledge-based generalization is contrasted by a limited number of actual implementations of expert systems in this area (e.g. Nickerson 1988a; Robinson and Zaltash 1989). Nickerson (1988a) has implemented a rather comprehensive method allowing the generalization of topologically structured linear features. The system offers a range of generalization operators (i.e. feature elimination, simplification, combination, and displacement), and is able to resolve interferences that may arise from previously executed generalization steps. It also handles the interrelated generalization of multiple feature classes (e.g. roads, rivers, etc.). It is probably the most operational and capable system besides the package reported by Leberl, Olson, and Lichtner (1985). Examples indicate that the method should be successful in generalizing linework of medium complexity (e.g. USGS digital line graphs, including hydrology, transportation, and political boundaries). Note, however, that such databases actually represent only a minor part of the complexity which may be found on common topographic maps.

Other implementations have focused primarily on algorithmic or engineering-like strategies. Some of the most advanced solutions that are applying an algorithmic approach are represented by the work of researchers of the 'Hannover school' (Powitz 1989; Meyer 1986; Lichtner 1979). Their research concentrates on the generalization of cadastral maps from 1 : 5000 to 1 : 25 000. They have developed a number of fine-tuned algorithms for the treatment of specific map elements (e.g. houses, roads, etc.). These solutions perform well for this narrow domain. However, as these techniques are focused on solutions for particular details, they are not general enough to cover more than just the specific tasks for which they were designed. At least some guiding knowledge at a higher conceptual level will, therefore, be necessary to control the application of individual algorithms to different situations.

Expert systems for map generalization

The application of AI and expert system strategies to the automation of map generalization represents a positivist approach. It is assumed that human knowledge and skill can indeed be formalized and modelled as a set of rules to automate eventually the behaviour of human experts. Great emphasis is put on knowledge. Expert systems derive their power from the knowledge they contain, not from the particular formalisms and inference schemes they employ. Or, as Waterman (1986: 7) put it, 'The accumulation and codification of knowledge is one of the most important aspects of an expert system.' Thus, successful formalization of knowledge (i.e. knowledge engineering) is crucial to the performance of such systems.

Common to all successful current expert systems is that they have addressed relatively well-defined problems (which is implied in the definition that expert systems 'attain high levels of performance in a narrow problem area'; Waterman 1986: 11). Examples of such domains are medical diagnosis, configuration of computer systems, and the like. Similarly in cartography, expert systems have focused on name placement (e.g. Doerschler and Freeman 1989) where a large body of knowledge (e.g. Imhof 1962) could be exploited, or, alternatively, map symbolization (e.g. Muller, Johnson, and Vanzella 1986) which could profit from theoretical research such as that by Bertin (1973). Apparently the success of expert systems strategies has been directly proportional to the availability of so-called public-domain knowledge (i.e. textbook knowledge). This is not to say that knowledge engineers have failed. However, textbook knowledge usually reflects the insight in a particular domain. For a given field, the less formalized is the knowledge contained in textbooks, the less advanced is the understanding of the problem (and the greater the effort which will have to be spent on knowledge engineering). Map generalization is one of those domains where little or no formal knowledge is offered by textbooks. There are many possible explanations why so few expert systems have been successfully implemented for this particular domain:

1. Expert systems are a relatively new technology to the GIS/cartography area and have so far not been addressed by many researchers.
2. Map generalization is a highly complex process. To name just a few of its characteristics: generalization partly encompasses intuitive and even artistic aspects; there are no clear criteria for performance evaluation (i.e. what is 'good' generalization?); it is, therefore, difficult to set up goal-driven approaches to generalization; generalization can take many forms, although some minimal formalisms must be satisfied (in order to ensure use of maps as a means of communication); and interdependencies between different constraints of generalization (i.e. map contents, scale, map purpose, targeted users, etc.) are highly convoluted.
3. Results from research in map perception tend to be rather specific (i.e. depending on the particular test set-up), and can only rarely be generalized (Spiess 1990). Moreover, meaning of map elements depends on a multitude of factors such as culture or economic objectives (Muller 1989).
4. Cartographers (i.e. experts on generalization) work in an intuitive, holistic fashion, and have problems decomposing their work process into a series of operations and steps. This is in contrast to the analytical approach used by experts in domains such as medicine, chemistry, and the like (i.e. the success areas of expert systems technology).
5. Cartographers are often reluctant to contribute knowledge to systems that clearly perform below their standards. For example, the relatively simple problem of cartographic line drawing (especially the drawing of dashed lines) is still not adequately handled by current cartographic systems. Cartographers would first like those deficiencies eliminated before pursuing further activities.

6. There is a mental gap between the work approach of knowledge engineers and that of the experts. On the one hand, knowledge engineers find it hard to understand cartographers since they have rarely worked in traditional cartography. On the other hand, cartographers find it hard to express their knowledge in terms of automated steps due to the reasons stated above.

Two insights can be gathered from the above discussion. It is clear that generalization cannot be solved through rigid, algorithmic methods, but has to be addressed by a combination of knowledge-based and algorithmic techniques. On the other hand, conventional expert systems and knowledge engineering methods for map generalization have not been overly successful and should be replaced by strategies that allow more structured knowledge acquisition. Expert systems cannot be expected to be improved far beyond their current state if knowledge acquisition is not improved. That is, it is doubtful whether the quantum leap anticipated by Muller (1989) will be achieved by using conventional knowledge engineering approaches.

An alternative approach: amplified intelligence

Amplified intelligence represents an alternative strategy for the automation of map generalization. The objectives of this approach are first to overcome the weaknesses of algorithmic approaches by incorporation of knowledge into generalization systems, and second to eliminate the deficiencies of knowledge engineering strategies by providing a structured approach to knowledge acquisition.

The strategy, in principle, is not completely novel. A multi-year programme which investigated the development and use of high-level tools for the augmentation of human experts' skills in various application domains has been carried out at the Stanford Research Institute (Engelbart 1962). This project has been extremely influential on the development of interactive mouse-driven user interfaces, most importantly those typified by the Apple desktop interface (Apple Computer 1987). There is also some similarity to decision support systems (e.g. Sprague and Carlson 1982). Lastly, Stormark and Bie (1980) have suggested that interactively controlled procedures should be used for cartographic generalization instead of mechanistic batch solutions. In addition to previous research along this line of thought, it is proposed here that amplified intelligence approaches can be utilized even further as a tool for increasing our understanding of complex design processes (i.e. for knowledge engineering).

Amplified intelligence is based on the idea that humans perform particularly well in tasks such as holistic reasoning, visual perception, design, and the like (i.e. those aspects of intelligence still withstanding formalization); computers, on the other hand, excel at solving repetitive, time-consuming tasks. In amplified intelligence the computer system solves a given problem in teamwork with a human expert. The operator initiates, controls, and evaluates system functions which the system performs automatically. The interactive work flow is supported by visual feedback (i.e. multiple views show source map, target map, and intermediate solutions suggested by the system). The system's operations may be accepted by the user or rejected and rerun using different parameters. These functions operate at a high conceptual level relating to design steps and units. That is, procedures work at the level of 'amalgamate selected buildings' or 'simplify selected lines to extent Y', rather than 'move point P' or 'delete line L'. Thus, these functions relate to generalization operators such as those proposed by Shea and McMaster (1989). Compared to current interactive graphics editors this will allow the cartographer to concentrate on design decisions rather than on detailed editing operations. Ideally, the desired map should be obtained through just a few operations.

An example, based on earlier research (Weibel 1989a, b) in the realm of automated terrain generalization, should clarify this approach. Although that research did not culminate in an amplified intelligence system as outlined below, components were developed that may be used in such a system. Table 10.1 shows a possible work flow by which a generalization of a terrain model may be obtained through a series of user actions and appropriate system operations. Note that in principle, every individual step in that sequence may be reinitiated based on an evaluation of its result. Figures 10.1 and 10.2 should help to illustrate this idea further.

The actual generalization operators can be based initially on existing algorithmic procedures that solve specific subproblems of generalization. Some of the existing algorithmic techniques for map generalization perform appropriately for their intended use. Their main problem, however, is that they employ no control knowledge, and attempt to implement generalization as a sequence of operational (and possibly mathematical) procedures in batch mode. As soon as control and design knowledge is brought into an interactive environment (through involvement of human experts), the performance of those algorithms can certainly be improved.

In order to make amplified intelligence strategies work a number of requirements have to be met. Systems have to be built that actually amplify the performance of experts. Experts must be offered the facilities to express freely their design ideas and decisions. System functionality and operation may best be compared to that of paintbox systems or powerful photo-retouching programs on advanced microcomputers and workstations. The following is a list of some prerequisites for amplified intelligence:

1. *Basic software:*
    (a) a powerful user interface based on the design paradigm of direct manipulation and visual feedback. Straightforward user interaction mechanisms, relating to manual designing, drafting, and retouching;
    (b) powerful visualization tools, including concurrent multi-window

Table 10.1 A possible sequence of operations for terrain generalization in an amplified intelligence context

User design decisions	System operations (automatic)
Start up application, select data window, set global parameters. Initiate generalizaton	
	Display statistical surface properties (relating to surface roughness)
Decide on thresholds for given map scale and purpose	
	Classify Digital Terrain Model (DTM) into relief types
Possibly modify and edit some classification errors	
Decide on generalization method (in our case, generalization based on structure lines is selected)	
	Extract network of structure lines (location, topology, and descriptive attributes)
Control degree of extraction, assess quality of result	
Decide on criteria and parameters for a 'select' operation. Initiate operation	
	Eliminate irrelevant links and update topology and attributes
	Highlight complex structure lines (the threshold for 'complex' is set globally in the initialization step)
Decide to carry out a 'simplify' operation on highlighted objects (possibly reselect), set parameters, initiate operations	
	Simplify ridges/drainage channels
	Highlight 'close' landforms (as candidates for combination)
Decide to carry out a 'combine' operation on highlighted objects (possibly reselect), set parameters, initiate operations	
	Combine those landforms
	Highlight other possible spatial conflicts
Set parameters for a 'displace' operation, initiate it	
	Perform feature displacement
Edit and retouch details	

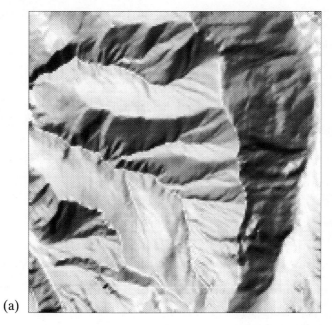

(a)

⊢————————⊣
1 km

(b)

⊢————————⊣
1 km

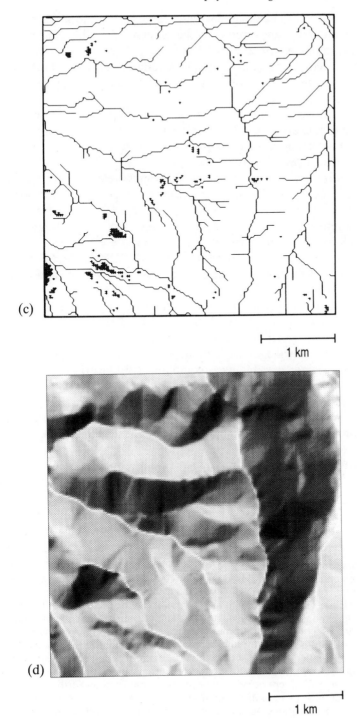

(c)

1 km

(d)

1 km

display of the source map/database, target map, intermediate results, and solutions suggested by the system. Possibly instantaneous transformations operated by sliders;

(c) raster–vector integration at graphics/design level: raster backdrop for vector images; full set of basic edit capabilities for vector, and paint functions for raster;

(d) basic cartographic functions performing at appropriate level: cartographic line drawing, cartographic text and symbol manipulations, colour selection and manipulation.

2. *Application-specific software:*

(a) generalization-specific functions which must relate to the conceptual level of a cartographer's design decisions (cf. generalization operators of Shea and McMaster 1989, or see Beard 1991, Ch. 7 this volume). Additionally, a set of functions for retouching and editing;

(b) the system should present suggestions: e.g. display statistical information on feature clustering, complexity, etc.; identify lines that are too complex; propose tolerances; identify overlaps and obscured objects, etc.;

(c) scripting mechanisms, along with transaction logging, in order to record and replay generalization work flows (e.g. to produce standardized generalizations in a production environment);

(d) tools for knowledge acquisition: mechanisms for operation/transaction logging; possibly even computer assistance in rule generation from transaction logs; a teaching mode to introduce new rules; identification of conflicts between rules, etc.

3. *Human knowledge:* skilled cartographers (experts) to control system operation (at least at an initial stage of system development).

Amplified intelligence tools for specific applications could be developed on the basis of these prerequisites. It is doubtful whether current cartographic systems can meet the above requirements. In a first step, improvements should concentrate on basic software functions (to form a solid basis for further development). If the approach of amplified

---

Fig. 10.1 Some steps of a possible work flow in terrain generalization, showing a selection operation. Based on an original terrain surface (a), structure lines (b) are automatically extracted to form the structure line model (SLM). The level of detail may be controlled by the user. On the basis of suggestions by the system and personal judgement, the human operator then initiates a generalization process. In this case, a selection operation eliminating all exterior links from the channel and the ridge network has been applied (c). The system then reconstructs a generalized surface model from the generalized SLM. Its visualization (d) lets the user decide whether to accept or reject that solution, and whether further generalization operations are necessary. An amplified intelligence approach requires short system response times, as well as concurrent visualization of the original data and intermediate products. DTM data courtesy of Swiss Federal Office of Topography

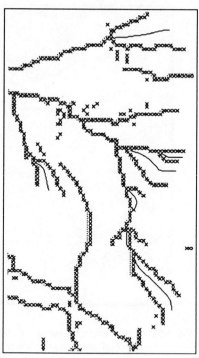

(c)

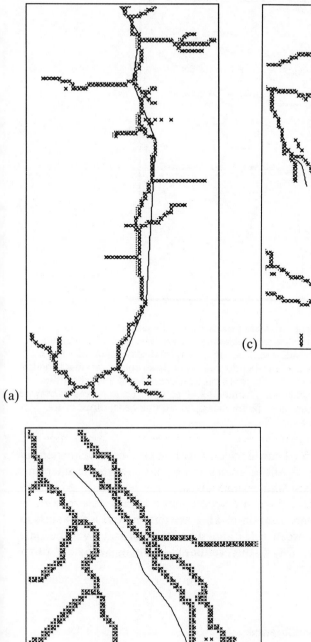

(a)

(b)

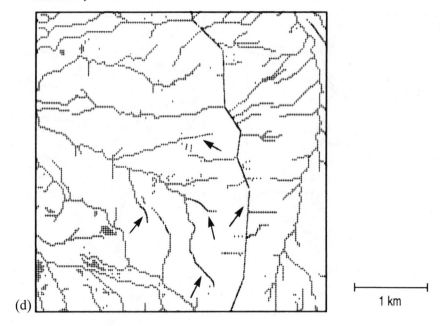

(d)

1 km

Fig. 10.2 In addition to selection, other processes may be used in terrain
generalizaton: (a) an example showing the simplification of a ridge line;
(b) combination of adjacent drainage lines; (c) displacement of a ridge
away from a drainage line; (d) the results of these operations in a broader
context (arrows indicate affected structure lines). Solid lines show the
location of generalized map elements. All generalization processes may be
applied either to an entire feature class, to a group of features, or to
individual features

intelligence is to be used for knowledge engineering and the development of
knowledge-based systems, it will be crucial to employ powerful mechanisms
for knowledge organization and evaluation (i.e. for learning). Amplified
intelligence systems work in close interaction with a human expert. Yet, as
soon as parts of the human decision-making are automated, issues such as
resolution of conflicts between rules and scheduling of rules will become
urgent (Armstrong 1991, Ch. 5 this volume) – just as with any other
knowledge-based approach.

Artificial vs amplified intelligence

This section tries to summarize important aspects of the amplified
intelligence approach, and compare them to the characteristics of expert
systems strategies. Of course amplified intelligence has some disadvantages:

1. Map generalization is not fully automated:
   (a) the system still depends on skilled labour (i.e. cartographers). Expert knowledge is not built into the system (however, it may gradually be built in to reduce the amount of human intervention required);
   (b) objective generalization is not possible using this strategy (however, is it at all possible, and desirable?);
   (c) the possibility that unskilled users produce bad generalizations is still not eliminated (it is hoped, however, that this situation will improve to some extent as better tools are developed).
2. Amplified intelligence may not be as attractive a research strategy as AI, since it is rather conservative, and it specifically addresses many problems which are technically oriented and are considered by many to be drudgery.

The above drawbacks are contrasted by a number of important advantages gained by using amplified intelligence:

1. Amplified intelligence is more conservative, i.e. less speculative than AI.
2. Knowledge is contributed by human experts in a direct way. Knowledge does not need to be formalized; even informal knowledge can be exploited.
3. Amplified intelligence may be based on existing solutions and, thus, be used for productive work immediately.
4. It may be useful as a research tool. Current generalization techniques can be evaluated and improved. Amplified intelligence systems provide a structured strategy for knowledge engineering.
5. It leaves creativity with the user to devote attention to interesting aspects of map production. Design decisions remain the user's responsibility. Tedious operations are automated. Acceptance of new technologies is improved, and increased motivation and work satisfaction are created (Stormark and Bie 1980).
6. Users may be motivated to explore different design alternatives. This way, more diverse and interesting-looking maps would be produced (Spiess 1990). This advantage can also be interpreted as a drawback (lack of objectiveness in generalization, see above).
7. Amplified intelligence may provide a method for prototyping an approach to AI.
8. This approach could also be used for other areas of map design (Weibel and Buttenfield 1988).

Although the amplified intelligence strategy has some disadvantages, it seems to be the only realistic method for successfully automating more complex tasks in map design and generalization. It is important to note that amplified intelligence pursues the same long-term objectives as AI. It is just a more conservative, evolutionary approach.

Summary

The application of expert systems strategies to map generalization has not yet been very successful. This lack of performance is mostly related to limited effectiveness of knowledge engineering. Amplified intelligence offers an alternative to overcome this principal weakness of AI. It still relies on human experts to solve problems, but provides them with tools that solve their problems faster and more satisfactorily, and thus amplifies their performance.

Amplified intelligence systems take an intermediate position between algorithmic and pure expert system strategies. They can initially be based on existing techniques (algorithmic or knowledge-based methods) that solve individual parts of generalization. The guiding knowledge ensuring proper application of those procedures is contributed by a human operator (expert). It is hoped that by analysing how cartographers use the system's functions, expert knowledge can be formalized in terms of automated methods and built into the system. Thus, such an approach may eventually lead to a full-scale expert system that entirely automates map generalization.

Future research will have to prove the feasibility of the proposed strategy based on implementations and empirical studies. It is suggested that initial development focuses on the design and implementation of amplified intelligence systems capable of handling generalization tasks that are relatively narrowly defined and well researched, such as feature selection, or line simplification and smoothing. These prototypes could serve as a basis for evaluating and developing specialized user interfaces and interaction procedures. The performance of the generalization results could be compared with the quality of corresponding algorithmic or rule-based solutions to estimate the potential performance increase. Lastly, the power of the proposed approach as a knowledge engineering technique could be assessed.

Acknowledgements

Robert Weibel completed this research while employed by Computervision GIS, Inc., Zurich, Switzerland. Travel funding from Computervision to attend the Syracuse Symposium is gratefully acknowledged. The views expressed in this chapter are those of the author, and do not necessarily represent those of Computervision GIS, Inc.

The experiments shown in Figs 10.1 and 10.2 were based on digital terrain data used by courtesy of the Swiss Federal Office of Topography, Wabern, Switzerland (permission granted 16 October 1990).

Part IV

Computational and representational issues

11

The role of interpolation in feature displacement

Mark Monmonier

Prologue

This presentation might appear less optimistic than most about the potential of artificial intelligence (AI) approaches to cartographic generalization. I must be careful, though, to acknowledge the distinction between the AI paradigm and the entity called a rule base, towards which this symposium is directed. I accept enthusiastically the notion that rule bases can have an important operational role in computer-assisted map design, yet I am highly sceptical of both the philosophical underpinnings of the AI paradigm and the extent to which complex AI and expert system strategies can be – and need be – employed to generalize maps.

This scepticism reflects in part the disenchantment of having naïvely embraced an attractive paradigm that could not easily be made to work. Early on, even before the terms 'artificial intelligence' and 'expert systems' were in vogue, I eagerly awaited the triumph of machine intelligence over the complex task of designing maps. In 1976, for example, at the Spring meeting of the American Congress on Surveying and Mapping, I confidently affirmed that 'computers can be trained to simulate the trial-and-error searching behavior of a human cartographer' (Monmonier 1976: 407). At that time much of my research focused on optimization algorithms, which not only worked but proved useful in dealing with a variety of problems in the design of choropleth maps and obliquely viewed statistical surfaces. Further successes nourished my enthusiasm for computerized intelligence, and in 1982, in the now obsolete textbook on computer-assisted cartography, I again identified with true believers in asserting that 'the similarities in switching behavior of the digital computer and the human nervous system suggest further that machines can be programmed to respond to a variety of complex situations in ways similar to the reactions of humans' (Monmonier 1982: 30). As Winograd and Flores (1987: 14–23) point out, the rationalistic tradition of science has encouraged this highly optimistic yet somewhat myopic search for context-independent rules for solving problems with mechanized logic.

Theorizing about solutions is, of course, easier than solving problems, and a failure to find ready answers often turns blind allegiance into wary agnosticism. My test of faith came in the mid 1980s, when I sought to generalize multiple linear features. Earlier in the decade my experiments with the raster-mode generalization of land use/land cover data demonstrated that polygon boundaries could be suitably generalized for presentation at a substantially reduced scale if the algorithm focused on their internal areas, rather than on the network of boundaries (Monmonier 1983). I wanted to extend this work to linear features such as roads and streams, and to cope with multiple features simultaneously. Then as now, most articles on generalization dealt with line features conveniently isolated from their neighbours and thus free from graphic interaction requiring displacement.

A solution that was both pragmatic and parsimonious seemed to lie in accepting the developmental limitations of algorithms and making the cartographic database more 'intelligent' by linking a more detailed representation of a map's features to a less detailed set of the same features. In 1986, I advocated this approach and concluded that 'intelligence in the data can obviate the need for highly elaborate algorithms'; I conceded somewhat dubiously that 'an expert-systems methodology might also be useful' (Monmonier 1986: 264). In this chapter I hope to refine this notion into something more than a precarious walk atop the paradigmatic fence. Graphic problems, after all, might benefit from geographical knowledge and cartographic expertise stored as graphic rules in a graphic format (Monmonier 1990).

Feature interpolation offers an implementation path for the hierarchical data model of Brassel (1985), who saw the advantages for generalization of an intelligent cartographic database with different levels of detail. This pragmatic approach is similar in spirit to the recognized need in knowledge engineering for generalization relations, which can guide users through an abstraction hierarchy (Parsaye *et al.* 1989: 418). None the less, a strategy that transfers most of the intellectual burden from the designer of the algorithm to the developer of the database can seem more expedient than pragmatic. Indeed, such a straightforward solution, which sidesteps more enigmatic challenges, might seem a form of computational cheating. As often happens, though, an apparent breakthrough in one part of the solution space merely highlights intimidating obstacles elsewhere. In this chapter I hope also to demonstrate that rule-base concepts have a definite role in making interpolated generalization work and in developing the required intelligent cartographic database.

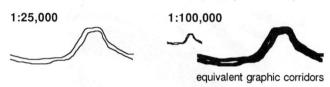

1:25,000   1:100,000

equivalent graphic corridors

Fig. 11.1   Parallel features intended for display at 1 : 25 000 (left) require graphic corridors four times wider when displayed at 1 : 100 000 (right)

Interpolated generalization and cartographic caricatures

A simple example demonstrates why feature displacement is the impetus for interpolated generalization. Figure 11.1 shows two roughly parallel features in a cartographic database intended for display at 1 : 25 000. If represented by lines 0.8 mm wide, these features occupy graphic corridors 20 m wide. When represented at 1 : 100 000 by similar line symbols, the features require graphic rights-of-way 80 m wide, which tend to overlap or crowd each other unless the centre lines of the features are shifted apart. Identifying areas of graphic impingement, computing displacement vectors, moving the features apart, smoothing alignments to avoid uncharacteristically abrupt jogs, and checking to ascertain that the lateral shifts have indeed avoided graphic interference. It is a complex, computationally demanding, iterative process, especially where several features share a common transportation corridor, as in the left half of Fig. 11.2. Johannsen (1973). Johannsen developed an early, interactive solution to computer-assisted feature displacement, and Christ (1978), who experimented with a quasi-satisfactory 'fully automated' approach, demonstrated that determining the direction and extent of displacement is not easy when the algorithm has access only to the single database to be generalized. Although Nickerson (1988a) more recently described an impressive algorithmic solution that works without manual intervention, a truly parsimonious approach can direct displacement using an existing smaller-scale cartographic database, as in the right half of Fig. 11.2. Small-scale representations with the features suitably separated for display at, say 1 : 500 000, provide 'target alignments' towards which the features can be moved for non-interfering display at a variety of intermediate scales.

If corresponding features in the large-scale and small-scale databases are divided into matched sections with linked endpoints, displacement becomes a computationally straightforward process of generating intepolated inter-mediate-scale feature alignments between these respective representations (Monmonier 1989b). Interpolation can proceed section by section, displacing first the endpoints and then repositioning the intermediate points in accord with each section's two displaced endpoints. As Fig. 11.3 demonstrates, an endpoint can be moved either to an interpolated position, midway between the corresponding endpoints of the large- and small-scale representations, or

Fig. 11.2 Portions of different maps enlarged to the same scale, approximately 1 : 10 000, to illustrate the displacement and smoothing of cartographic generalization. Closely spaced parallel features sharing a transportation corridor (below) would readily suffer graphic interference if displayed without displacement at a substantially reduced scale. Small-scale representations (above) offer displaced alignments towards which the features can be displaced for display at a variety of intermediate scales. Thick black lines in the upper panel represent the railway and principal highway and the thinner bifurcating line between them is the creek represented in the lower panel by separate lines for opposite banks. (*Sources:* USGS, Bellefonte, Pa, 7.5-minute topographic map, 1 : 24 000, 1962 (below). USGS, Harrisburg, Pa sheet, 1 : 250 000 topographic series, 1962 (above). Examples show part of the water gap of Spring Creek, north of Bellefonte, Pennsylvania.)

to an extrapolated position, along the line through the endpoints but beyond the small-scale representation. A test for graphic interference is needed to determine whether an initial displacement based upon the sizes of the features' graphic corridors is adequate. The degree of displacement can be adjusted accordingly, and if necessary, extrapolation can move the feature past its small-scale representation.

Figure 11.4 illustrates two strategies for displacing the section's intermediate points. For sections with little curvature, the separate proportional translation and rescaling of the X and Y axes can effect a suitable two-

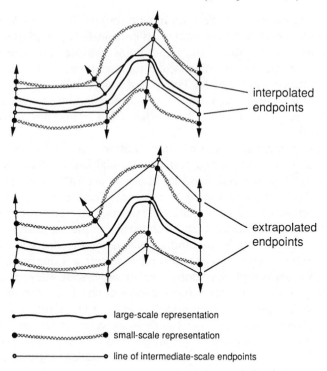

large-scale representation

small-scale representation

line of intermediate-scale endpoints

Fig. 11.3 Endpoints for features' displaced intermediate-scale positions can be interpolated between the large- and small-scale representations (above) or extrapolated beyond the small-scale representation (below)

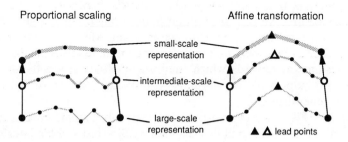

Fig. 11.4 Proportional scaling (left) and affine transformation (right) provide for the displacement of points lying between a section's endpoints. Proportional scaling is reliable only for relatively straight sections, whereas an affine transformation can specify interpolated positions for the intermediate points of sections with a single prominent bend. In addition to the endpoints, the affine-transformation strategy requires matched lead points on all three representations

dimensional transformation of the intermediate points. For sections with a single pronounced bend, a six-parameter affine transformation based upon the two endpoints and a designated lead point provides an appropriate transformation for the intermediate-scale representation. Corresponding lead points identified on both the large- and small-scale representations supplement the endpoints in capturing the essential curvature of the section. Features should be regionalized so that each section has no more than one pronounced bend.

Interpolated generalization supports two additional generalization operators: simplification and smoothing. Caricature-weighted simplification, for example, modifies the Douglas line-simplification algorithm (Douglas and Peucker 1973) by increasing the likelihood of selecting points near the caricature and eliminating points relatively far from the salient curves and exaggerated protrusions captured and enhanced by the caricature (Monmonier 1989a). Interpolated smoothing, which would precede feature displacement, operates section by section to: (1) project a duplicate of the caricature or small-scale representation on to the large-scale feature; and (2) shift each point in the large-scale representation closer to that duplicate (Monmonier 1989b: 53–4). This approach, which assumes that the small-scale representation is smoother than its large-scale counterpart, moves each point along a line perpendicular to the duplicated caricature and uses the relative scales of the large-, intermediate-, and small-scale representations to determine the amount of movement.

Although discussed heretofore solely in the context of line generalization, cartographic caricatures can also assist in the interpolated generalization of point features, area features, and map labels. Moreover, a broader concept of the cartographic caricature would support not only the smooth, continuous transition of a symbol between its large-scale and small-scale positions as the map scale is decreased but also the abrupt shift of features, as when generalization must merge the opposite banks of a river into a single linear feature. Thus a single point feature might have large-scale and small-scale locations that define a trajectory along which its symbol can be displaced, whereas a group of two or more point features might have a common destination towards which they might either gravitate for minor reductions in map scale or abruptly consolidate for more substantial amounts of generalization.

Similarly, as Fig. 11.5 demonstrates, a map label might have a narrow range of scales for which small displacements can avoid graphic interference as well as a new, more distant, position from which a leader line links the label with its associated symbol. Although line generalization might accommodate reductions in scale for large area features by operating on their boundaries, smaller features, such as groups of small islands or patches of forest land separated from a larger nearby preserve, might require cartographic caricatures that represent amalgamation and other abrupt transitions. Moreover, generalization sometimes alters the dimensionality of a cartographic feature. As Fig. 11.6 illustrates, a city represented as an area

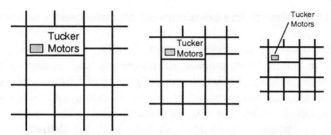

Fig. 11.5 For a small range of scale, minor amounts of displacement might maintain an aesthetically acceptable separation of a label from its associated symbol and other nearby features (left and centre). Smaller scales might require that the label be shifted to another part of the map and linked to its symbol by a leader line (right)

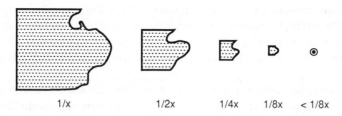

Fig. 11.6 Progressive reduction in the scale of an area feature such as a city might eventually require its treatment as a point feature

feature at a range of intermediate scales might collapse to a point feature at much smaller scales. An automated display system able to generalize maps across a wide range of scales might require not one but several caricatures for each feature.

Cartographic caricatures most certainly constitute a significant and potentially complex cartographic knowledge base. Indeed, the effectiveness of interpolated generalization in remediating graphic interference and providing appropriate visual clues for a simplified, smoother geographical frame of reference depends upon the geographical knowledge captured in the small-scale representation. In guiding interpolated generalization, cartographic caricatures serve as graphic rules, appropriately encoded to promote graphic solutions to graphic problems. Unlike tabular or algorithmic rule bases presumed to provide location-independent solutions, cartographic caricatures are idiographic rules, tailored to specific places and specific features – less enigmatic perhaps than more universal drafting rules but fully as important. If digital cartographic data are to sustain map display over a broad range of scales, cartographic caricatures linked to their larger-scale representations might become as pragmatically essential as topological structure and the relational database model.

Rule-base strategies for the development of cartographic caricatures

Conventional rule-base strategies (Shea 1991, Ch. 1 this volume) can be useful both in adding cartographic caricatures to an existing large-scale database and in enabling this intelligent geographical database to address a variety of locationally independent problems in cartographic display. This section examines the roles that rule bases might play in the development of cartographic caricatures. These roles include both the delineation of the caricatures themselves and the joint, matched regionalization of large- and small-scale representations of features.

Matched regionalization requires a varied rule base able to support key types of decisions. Rules are needed: (1) to determine correspondence among representations on the basis of similarity in location, shape, and topology; (2) to match nodes at which various features intersect; (3) to detect exchange crossings and their endpoints; (4) to limit section curvature so that an affine transformation can direct the interpolation of intermediate points of single-curve sections; (5) to direct the order of application of these and other rules; and (6) to identify unresolved situations requiring human intervention.

Computer-assisted matching of features in existing databases covering the same region at different scales must depend largely upon equivalent feature codes and similar geographical positions. But angles, directions, general alignments, and patterns of linkages with similar and dissimilar features also provide useful clues. The matching process is thus akin to the early stages of conflation (Saalfeld 1988), which matches features by exploiting similarity in shape and topology to compensate for inexact geographic correspondence of equivalent representations and vague or missing labels. Indeed, matching features and endpoints in different databases would be much more straightforward if the attribute lists of all linear features included names as well as codes merely assigning features to general categories. Yet because large-scale databases commonly fail to identify the various chains or arcs representing a single motor route, such as 'Erie Boulevard' or 'New York State highway 92', the rule base must deal with the possibility of abrupt changes in direction at intersections.

Since interpolated generalization requires not only matched features but matched endpoints, identification of corresponding intersections at different scales is an essential part of the process. But matching is confounded because many intersections in the large-scale database will lack counterparts in the small-scale database. These unmatched intersections will be important skeletal elements when, at a later stage, caricatures missing in the small-scale database are inserted or generated. Yet because the small-scale database omits some features, the rule base must recognize that the convergence of six roads at 1 : 24 000 might well correspond to the convergence of only four roads at 1 : 250 000, as in the upper part of Fig. 11.7. Moreover, a six-way intersection at 1 : 250 000 can represent two close four-way intersections at 1 : 24 000, as in the lower part of Fig. 11.7. The

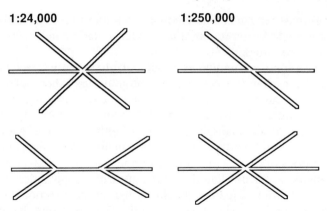

Fig. 11.7 Hypothetical examples of how the omission and merging the features for portrayal at smaller scales can alter the number of roads converging at an intersection

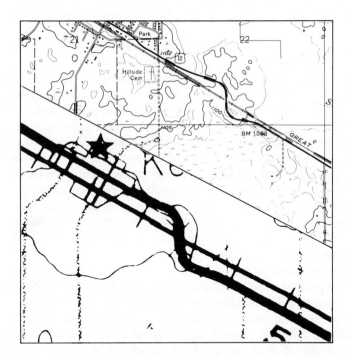

Fig. 11.8 Comparison of an exchange crossing represented at 1 : 24 000 (above) and 1 : 250 000 (below). Smaller-scale portrayal has been enlarged to roughly the same scale as the large-scale representation. Examples show an area just south-east of Kerkhoven, Minnesota, where highway US 12 crosses the tracks of the Great Northern Railway, now a part of the Burlington Northern Railway. (*Sources:* USGS, Kerkhoven, Minn., 7.5-minute topographic map, 1 : 24 000, 1979 (top). USGS, St. Cloud, Minnesota, 1 : 250 000 scale series, 1979 (bottom).)

rule base thus must recognize that comparatively short sections in the large-scale database might be represented at smaller scales by null sections, that is, sections of zero length.

The rule base must accommodate a number of special intersections, one of which is the exchange crossing that often occurs where two parallel features such as a railway and a highway share a common alignment for a considerable distance. Occasionally the 'subordinate feature', typically the highway because it was built later, will abruptly cross the path of the 'dominant feature' and continue along the common alignment but on the dominant feature's opposite side. Although the point of cross-over forms a convenient endpoint for dividing the dominant feature, the entire cross-over portion of the subordinate feature should form a separate section connecting endpoints at which the subordinate feature diverges from the common curvature of the two parallel features. As the example in Fig. 11.8 illustrates, in smaller-scale representations cross-over sections are often attenuated in the direction perpendicular to the dominant feature. An earlier study examined several measurements and their threshold values that can be included in a rule base for detecting exchange crossings (Monmonier 1989c).

Rules are also needed to limit the curvature of sections, particularly in the large-scale database, so that a six-point affine transformation based on the endpoints and matched lead points can adequately transform the section's intermediate points. The maximum curvature rule, which need be applied only between matched nodes, might be based recursively upon the maximum perpendicular deviation from a straight line between recognized endpoints, in a manner similar to the bandwidth criterion of the Douglas line-simplification algorithm (Douglas and Peucker 1973). A similar rule or set of rules might identify matched lead points, and a related rule can identify sections with minimal curvature, for which proportional rescaling can interpolate intermediate points and which thus do not require lead points. For a long sinuous feature not interrupted by intersections with other features, marked changes in trend direction afford another criterion for linear regionalization (Buttenfield 1987).

Although an existing small-scale cartographic database might usefully be regionalized and matched, feature by feature, with a large-scale database to support interpolated generalization, specially prepared caricatures will be required for features without corresponding representations in the small-scale database. Because features inserted later in the process would tend to be more contorted than those added earlier, in order to avoid graphic interference, a rule base might promote consistency by specifying the order in which various types of features should be caricatured manually. This rule base might also monitor the work of the geographer crafting the caricatures, and not only determine matched endpoints for the caricatures but also suggest the need for manual adjustment in portions of the map where, for example, a tightly curved feature might overlap itself at a substantially reduced scale. A line-generalization algorithm operating on the large-scale database might even suggest trial caricatures by generating simplified, smoothed representations (Buttenfield 1991, Ch. 9 this volume). Subsequent analysis based on a set of rules might then detect areas with tight curves or close features requiring manual adjustment.

Rule-base strategies to assist caricature-guided cartographic generalization

An operational model of interpolated generalization requires rule bases to resolve the graphic conflicts that occur when several symbols compete for the same space on the map. Although idiographic rules coded as cartographic caricatures provide both expedient displacement and geographically meaningful simplification and smoothing, a more universal set of display rules might (1) specify when extrapolated displacement is excessively distorting the shapes of graphically congested features and (2) select those features to be omitted. In some cases the rule base might even call for less graphically demanding symbols, such as an abbreviated spelling of a name or the merging of opposite riverbanks into a single line. Moreover, because the map author must be able to adapt the reduced-scale display to unique analytical requirements or communications objectives that the expert who designed the caricature could not anticipate, an efficient and effective automated system for cartographic generalization needs modifiable rules that can be tailored to the map's goals and applied consistently to the entire display.

Rules to assist interpolated generalization should be based on fundamental generalization operators such as those in the model of cartographic generalization developed by McMaster (1989a). Since interpolated generalization deals directly with features represented in vector mode using plane coordinates, the relevant fundamental generalization operators for line features are omission, simplification, smoothing, displacement, and merging, whereas those applied to point features are omission, aggregation, and displacement. McMaster's list includes enhancement as a basic operator for line features, but the enhancement operator is not relevant here because generation of an intermediate-scale display through interpolation would rarely if ever require cartographic detail not already existing in the large-scale representation. Although a caricature might exaggerate or 'enhance' the prominence of features important as visual clues, this type of exaggeration involves the collaborative effects of simplification, smoothing, and displacement.

Close inspection and comparison of existing maps representing the same areas at different scales suggest that cartographers using non-automated methods of generalization often combine smoothing, simplification, and displacement into two composite operators: one that largely alters shape in the process of moving features apart to avoid graphic interference and a second that amplifies some shapes or arrangements of features to overstate geographically significant details. (At the risk of adding unnecessarily to cartographic generalization's rapidly expanding lexicon, let me suggest the names **accommodation** for the former and **embellishment** for the latter.)

Although straightforward smoothing and simplification often occur in non-congested parts of the map, cartographers seem to rely upon five basic strategies for dealing with graphic interference: drop it, move it, merge it, exaggerate it, or substitute a symbol with a different geometry, as when a circle or diamond replaces a complex clover-leaf intersection on a limited-access highway. Automated generalization based on the interpolation concept might need to adopt a similar strategy by closely linking the processes of smoothing and simplification (McMaster 1989b) or by co-ordinating the application to individual features and pairs of features of carefully chosen sequences of operators (Monmonier and McMaster 1990).

Prevailing generalization practices at government mapping agencies (McMaster 1991, Nickerson 1991, Chs 2 and 3 this volume) would seem to offer few rules suitable for interpolated generalization. Scale reduction is substantial between map series only a rung apart on the ladder of scales, and the most prevalent and obvious generalization operator is omission (Mark 1991, Ch. 6 this volume). Eliminating entire coverages or feature classes is easier, after all, and probably less controversial than the careful development of display priorities to accommodate a range of general users with varied interests. If anything, published maps suggest that the rules in use are too broad, too rigid, and too insensitive to the local importance of features in sparsely occupied sections of the map.

Consider, for example, the relationship between the US Geological Survey (USGS)'s 1 : 24 000 scale, 7.5-minute topographic series and its 1 : 100 000 scale, 30 × 60 minute planimetric series. Omitting contours avoids some graphic congestion on the smaller-scale map, yet only serves to highlight the transportation fetish that must have guided selection of the remaining features. USGS also produces 1 : 100 000 scale quadrangle maps in a topographic–bathymetric edition, essentially the planimetric edition with the addition of contour lines and a light-brown tint to represent strip-mined areas and other distorted surfaces. The smaller-scale contour plate is generalized principally through use of a larger contour interval but also through smoothing, simplification, and the merging or omission of small outliers.

The 1 : 100 000 scale map justifiably preserves principal roads, active railways, and inland waterways, and thin but legible line symbols permit the unobtrusive yet detailed retention of all secondary roads, city streets, and intricate interchanges. (That so much of the highway network can be retained suggests that American large-scale topographic maps, in contrast to their European counterparts, are rich in terrain but poor in culture.) Except for major landmarks, individual buildings disappear, and the grey tint representing built-up areas at 1 : 100 000 expands quite reasonably beyond the red-tint zones representing built-up areas at 1 : 24 000. Although what is important and worth displaying is, of course, a matter of judgement, the remaining features shown at the smaller scale often reflect what Emerson so perceptively called 'foolish consistency'. For instance, in its predilection for minor airstrips, 'old railway grades', and stream gauging stations, the USGS has perhaps inadvertently tailored the 1 : 100 000 scale series to a loose constituency of drug smugglers, railway historians, and hydrologists. This all-or-nothing strategy and its wholesale slaughter of entire feature categories is an expedient that an efficient and effective rule base for interpolated generalization might at least adjust to the specific goals of published maps or to the unique interests of individual analysts.

Comparison of the 1 : 100 000 scale and 1 : 250 000 series, fractionally closer than any of the USGS's four current nation-wide scales, also offers little guidance. Small towns and complex highway interchanges collapse to point symbols, shorelines and drainage channels suffer further smoothing, and many secondary roads and most residential streets disappear altogether, along with gauging stations, most landing strips, and even some abandoned rail lines. At 1 : 2 000 000, the map would in principle retain only 0.0144 per cent of the features and details found at 1 : 24 000 – and it probably retains a bit less.

What kind of rule base is needed, then, to support interpolated generalization? The solution seems to lie not in a large number of intricate rules but in a comprehensive set of display priorities to be implemented using a comparatively small set of universal rules. Clearly a finer, more detailed set of display priorities is needed, to provide a continuum of cartographic solutions tailored to an unlimited range of intermediate scales. These display priorities, which need to address both selection and displacement, should support cartographic customized displays for both individual users (Weibel 1991, Ch. 10 this volume) and specific communication objectives (Mackaness 1991, Ch. 13 this volume). Since many generalized maps will be displayed rapidly, on demand to serve specific applications, a satisfactory solution requires flexibility, not a rigid universal optimum.

Epilogue

Exploratory cartographic analysis requires a rapid, efficient process for generalizing geographical data across a range of display scales. Environmental scientists, epidemiologists, and other exploratory users of spatial data need to navigate quickly and freely through a cartographic database and to change scale in seconds, not minutes or hours. Highly interactive cartographic displays able to zoom in or zoom out efficiently require a balanced generalization strategy relying upon both universal rules and detailed local knowledge of the significance of geographical features (Richardson and Muller 1991, Ch. 8 this volume) and their graphic interaction. Satisfactory real-time performance calls not only for rule bases, expert systems, and improved algorithms but for intelligent databases.

An intelligent cartographic database will unquestionably be a big database. A rich set of feature codes describing the function, significance, and meaning of geographical features will permit universal rules to tailor the selection and omission of features to a user profile (Daniels 1986) or to the goals of the analysis. User profiles might also support a periodic or continual modification of both the display priorities and the graphic rules. If the system allows users to override default display priorities, the frequency of these overrides and the type of adjustment are new knowledge useful for adjusting the generalization rule base. Moreover, if the system permits users to revise the caricatures, these modifications might also lead to more effective graphic rules. Knowledge changes as experience accumulates, and a record of user behaviour can guide the growth towards maturity of an intelligent database.

For these reasons, research on cartographic generalization will become not just a display problem but an interface question, tied closely to work on user strategies (Beard 1991, Ch. 7 this volume) and query languages. An intelligent cartographic database will also be a highly integrated database (Armstrong 1991, Ch. 5 this volume), with a variety of relations or pointers linking features and sections of features to their larger- and smaller-scale representations as well as linking parts of features to other nearby features and to nearby parts of the same feature. Graphic rules suggesting displacement directions and smoothed and simplified forms will also be an important part of an intelligent cartographic database. Real-time implementation of interpolated generalization or a similar strategy calls for new machine architectures (Langran 1991, Ch. 12 this volume) adept at dealing with graphic rules and massive amounts of tightly related, hierarchical data.

Implementation of an interpolated generalization strategy for automated map generalization calls for considerable experimentation to guide the selection of a minimal yet effective number of small-scale and intermediate-scale representations. In an advanced implementation, for instance, some features might require two cartographic caricatures, one quite simplified and the other with certain elements exaggerated. The minimalist caricature would support generation of a visually recessive representation if the feature is to be part of a thematic map's background information, whereas the exaggerated caricature could support a more visually expressive representation when the feature must play a more prominent, foreground role.

As Armstrong (1991, Ch. 5 this volume) notes, generalization can require three types of knowledge: geometric, structural, and procedural. Since structural knowledge might recognize more than one function or level of meaning for a feature, multiple geometric representations might be needed, as well as the procedural knowledge to choose among them. This procedural knowledge might, in turn, include weights or priorities that guide the automated generation of reduced-scale representations with a blend of more than one caricature. The potential for needless complexity is enormous, however, and experimentation with typical users must address whether such computationally demanding conceptual embellishments make any subtantial difference in how viewers visualize what the map represents and how such different visualizations affect understanding and behaviour.

Acknowledgement

The partial support of the National Science Foundation (grant SES-86-17459) is gratefully appreciated.

12

Generalization and parallel computation

Gail Langran

Introduction

A useful exercise in creative thinking directs the thinker to imagine the world if one or more of its fundamental characteristics were altered. For example, how would our housing be designed if the earth had no gravitational field? How would our society differ if we had no written language? This exercise translates readily to a problem-solving environment: the problem-solver imagines his or her problem if one of its basic constraints were removed.

Many constraints combine to inhibit full automation of cartographic procedures. Constraints can be classed as geographical, technical, and conceptual. Geographical constraints relate to the irregularities and idiosyncrasies of the natural world and the difficulty of developing structures and logic to treat them. Technical constraints relate to limitations in computational performance and storage capacities. Finally, conceptual constraints occur when a researcher's outlook on a problem confines his or her approach to limited avenues. Arguably, while technology and innovation work to remove geographical and technical constraints, each individual must seek to lift his or her own conceptual constraints.

The Von Neumann bottleneck

Among the fundamental assumptions of algorithm design today is that the Von Neumann software model will be used. A software model dictates a particular approach to writing a computer program. As described by Backus (1978), the Von Neumann model involves a single processor, a data store, and a tube that transfers data between the processor and store in small units. Backus terms this tube the 'Von Neumann bottleneck', for obvious reasons.

This constrained outlook on software design inevitably limits innovation. The vast majority of the programming languages and computer architectures used today are built on the Von Neumann model and, unfortunately, a self-perpetuating cycle exists. Practitioners use Von Neumann languages and architectures because these are what the industry offers, and the industry develops new languages and architectures along the Von Neumann model because that is what practitioners use.

Benefits of alternative approaches

Without dismissing the Von Neumann software model, one can say that it would be poor practice to accept it as the only model without exploring others. Gelernter (1987) argues that a new way of thinking can introduce new ways of solving problems. Backus (1978) states that the Von Neumann bottleneck is also an intellectual bottleneck that has kept us tied to word-at-a-time thinking. If programming languages are expressions of software models, then development of programming languages should follow development of software models that are appropriate to specialized problems. Instead, there is a tendency to use passively the software models implemented in existing programming languages.

Spatial problems are excellent candidates for alternate software models because multiple dimensions pose ordering problems for sequential processes and because the generally large volumes of spatial data exacerbate the Von Neumann bottleneck. Parallelism offers apparent benefits for these reasons, and presents a reasonable basis for developing alternate models. Cartographic generalization appears especially suited to parallel models, since the holistic, highly visual manual process is difficult to replicate by standard (i.e. sequential) programming methods.

The discussion that follows examines a collection of parallel software models and explores their usefulness to generalization. It is not a treatise on implementing parallel software models for cartographic ends, it does not attempt to review the plethora of parallel computing architectures and programming languages, nor does it advocate one approach over another. Rather, the goal is to consider different ways to think about cartographic problems using generalization as an example, and to explore the benefits that alternative software models can offer to cartographic automation.

Order of discussion

The discussion that follows examines the components of generalization and considers what aspects of the task could be performed in parallel. It proceeds with a review of the problems and choices inherent in parallelism, then evaluates several alternate parallel software models and how each could be used in cartographic generalization. The final remarks are directed at the question of how cartographers and spatial information specialists can capitalize on parallel software models, and what directions future research might take.

Components of generalization

Generalization is comprised of component processes, although the precise nature of the components is a source of some disagreement. At the most basic level, generalization includes selection of the data to be depicted, simplification of its geometric rendering, and displacement of the graphics to resolve conflict (McMaster 1989a elaborates on these components of generalization). Rhind (1973) itemizes five mechanics of generalization: reduction of line sinuosity, transposition of features, amalgamation of features within and between categories, and changes to symbol classes. Beard (1987) divides generalization functions into five classes: reduction of line sinuosity, selection, aggregation, collapsing of a dimension, and coarsening of detail. Within this volume, several authors present additional classification schemes.

Regardless of how generalization is decomposed, most efforts to automate the generalization process would seem to assume that its components are discrete and relatively unrelated, and that a complete generalization procedure is no more than the sum of its parts, which are applied one after the other. Exceptions to this generality do exist; a recent example is McMaster's (1989b) study that examines how two generalization algorithms interact when applied together.

The interrelatedness of generalization procedures has long been known. Pannekoek (1962) observes that when a river has been simplified, it creates space on a map for a small settlement that otherwise would have been omitted. Conversely, when the small settlement must be shown, it may be necessary to simplify the surrounding features to create space. Selection, too, affects rendition; for example, when a road or stream is included in steep terrain, the contours must be adjusted accordingly. And the selection and rendition of features on maps affect displacement needs. Even assuming only three basic components of generalization, their interdependency is apparent. In Chapter 13 in this volume, Mackaness (1991) coins the term 'topogeneralization' to describe these interdependencies and demonstrates how different components of generalization are used to varying degrees based on the composition and juxtaposition of map elements.

Admittedly, a conventional (sequential) approach to generalization does produce acceptable results in many cases, although some interactive postprocessing is always assumed. However, if one were to attempt to model generalization procedures on the methods used by a human, then a sequential treatment would become unwieldy. Human vision provides a cartographer with a Gestalt view of the map to be generalized, and intuition and insight permit the cartographer to integrate the components of generalization (see Beard 1991, Ch. 7 this volume). Even without a full understanding of how information is processed within the human brain, parallel rather than sequential models of information processing seem more likely to mimic the human decision process. In fact, regardless of whether the human model is a desirable one, the parallel model appears to offer major benefits to generalization and other visual information processing problems simply because of the amounts of highly correlated information to be processed.

How to intertwine the components of generalization in our algorithms is problematic. A procedure that achieves that end could make the logic of separate selection, simplification, and displacement rules seem straightforward by comparison. Combining the rules and incorporating interdependencies could result in deeply nested and context-sensitive logic. Beard's constraint-based approach (Ch. 7 this volume) could alleviate the complexity of the 'rule-based monster'; dividing it could also aid in conquering it.

The question, then, is how to perform generalization in a holistic manner that mimics the manual process without producing a logic bottleneck that is potentially as formidable as Backus's Von Neumann bottleneck? A first step is to examine ways to decompose the generalization process. This is not unlike the problem of 'finding the objects' in object-oriented programming, which is discussed by Mark (1991, Ch. 6 this volume). However, an additional step beyond that needed in an object-oriented environment is to consider which units can or should be processed concurrently.

Component processes

The obvious choice for concurrent processing is to divide processing according to the component processes of generalization described earlier. Performing component processes in parallel would be a step towards integrating them if the results of computations were communicated between processors with sufficient frequency.

Spatial units

Because generalization is an inherently spatial process, we could generalize different units of space or different geographical features concurrently. If not enough processors were available to allot a processor to a unit, units could be processed in groupings, moving hierarchically, by band sweep, or radially. Two problems exist with this approach, however: spatial decomposition alone would not desegregate the functional components of generalization, and the boundary effect of processing spatial units separately would need to be overcome.

Thematic units

Different types of geographical information require different treatments. For example, single-point features do not need simplification, and features whose position is critical are not displaced. It may be reasonable to parcel out the logic and the data among processors according to thematic distinctions.

Procedural units

Inevitably, the computational process decomposes into segments that could be treated in parallel. For example, rules or subroutines whose outcomes are independent of one another could be executed in parallel to speed processing or improve its results. Richardson and Muller (1991, Ch. 8 this volume) describe the mutual importance of rules and procedures in a generalization system; theoretically, rules and procedures could reside on separate processors that are geared to the differing computational requirements.

In sum, the functional, spatial, thematic, and procedural portions of generalization are all reasonable candidates to process in parallel, although each provides slightly different benefits. Decomposing by procedural unit could improve performance, but it does not necessarily open doors to new directions. Decomposing by spatial unit could change data structures and logic drastically, but would not alone bring us closer to the immediate goal of integrating the subtasks of generalization. The next step, then, is to examine existing parallel software models and to explore implementation options. However, a brief introduction to parallel computing improves later discussions.

Concepts in parallel computing

Parallel computing can depart considerably from standard sequential methods. While it offers potentially improved performance and opportunities for new approaches, these benefits come at the cost of a new set of concerns for hardware and software designers. Parallelism assumes concurrent processes, which compete for memory or communication links unless formal procedures exist for resolving contention. To support parallelism, some overhead procedures become necessary that would be unnecessary without the parallelism. One example of this overhead is a set of protocols to resolve the contention for memory or communication links described immediately above.

Ideally, a parallel model minimizes idle machine parts. A computer could be idle as it awaits data, or data transport could stall as it waits for processing to be completed. The preferred computational mode is for all parts to be productive most of the time. An interesting, and often desirable, characteristic of software models is scalability, i.e. if more processors are added, performance improves. Arguably, scalable models are more versatile and upwardly mobile, since they can capitalize on reductions in processor costs.

Hardware design options

In addition to introducing new concerns, parallelism presents new options for information processing. Options tend to be associated with either hardware or software design, although often the two are linked. Hardware design choices dictate how many processors, how specialized the processors, and how to interconnect them. The options that follow are particularly important to these decisions.

Hardware can be designed to support a multiple instruction stream and multiple data stream (MIMD), or a single instruction stream and multiple data stream (SIMD) (Flynn 1972). A MIMD system provides a unique set of instructions to each processor by splitting the program into tasks and distributing the tasks among processors. In contrast, a SIMD system broadcasts a single set of instructions to all the processors.

Granularity refers to the (adjustable) number and power of the processors used. A coarse-grained system has fewer, more powerful, processors; a fine-grained system has more, but simpler, processors. Coarse-grained systems are associated with the MIMD approach, while fine-grained systems are associated with the SIMD approach. All processors in a shared memory system have access to a common set of memory. A distributed memory system permits each processor to control a portion of memory.

Communication can be fixed or general. A fixed communication machine provides specific channels of communication via hardware. Common network configurations include the ring, grid, hypercube, and binary tree. General communication machines permit any processor to communicate with any other, although some may have speedier access to others because of the pattern of wires and cables.

Software design options

Parallel machine architectures are seldom independent of a software model. In turn, parallel software models tend to be logical extensions of the parallel hardware upon which they are implemented. Software design choices dictate where tasks are allocated, how instructions are passed, and how data are stored and accessed. The software design options that underlie any parallel implementation include the method of task decomposition, the method for control of operation, and the heterogeneity of instructions passed through the system.

Domain decomposition divides a job into small tasks that are similar to one another (e.g. using spatial components), and allocates each task to a processor. Functional decomposition divides a job into tasks that differ from one another (e.g. using mechanical or thematic components), then sends the tasks to processors that specialize in a given class of task. An application can use both types of decomposition by functionally decomposing a job, then subdividing the subdivisions before parcelling them out to processors.

Control parallel algorithms are designed to have multiple threads of control (e.g. multiple instruction sets) that emanate from different processors. Data parallel algorithms use a single instruction set to modify a large data set simultaneously (Hillis and Steele 1986). This distinction is the software equivalent of the MIMD/SIMD dichotomy.

A heterogeneous system allocates different pieces of the job to different processors, which may have specialized capabilities. In general, a heterogeneous system is coarse-grained, uses a MIMD protocol, and uses a network of communications to pass information between processors regarding the results of work. The amount of overhead directly relates to the number of messages that must be passed between processors. A homogeneous system issues identical instructions to all processors, which execute the instructions independently. Homogeneous systems tend to be fine-grained and use SIMD protocols.

It is crucial to understand the variety of parallel approaches. Without this understanding, one could not appreciate the wealth of options available to software designers. Designers are relatively free to pursue their preferences, since no single parallel paradigm is universally accepted as superior and profound disagreement exists among information scientists concerning what a parallel program should look like. None the less, each approach offers unique capabilities that can be evaluated with respect to a particular task – in this case, that of cartographic generalization.

Parallel software models and generalization

Software models are the result of combining different software design options into logical and useful forms. The classification of models presented below was developed by Gelernter (1987). This section evaluates each model and considers how it could be implemented for cartographic generalization using the functional, spatial, thematic, or procedural decompositions of generalization presented in earlier sections.

Parallelization

It is possible to take a program written to run sequentially and compile it so it will run in parallel. While this is a marginally parallel approach and can scarcely be called a parallel model, it deserves mention because it is an effective and thriving means of exploiting parallel computers. Two versions of this method exist. First, a translation program can actually rewrite program code to exploit inherent parallelism and perform independent tasks concurrently. Extensions to FORTRAN are available that convert:

Do A, then B, then C

to

Do A, B, and C at the same time

The translation program identifies independent tasks based on which tasks do not depend on the results of a previous computation. Once identified, independent tasks are assigned to different processors. This approach is particularly valuable in solving the 'dusty deck' problem, where rewriting existing software to use a truly parallel software model would not be cost-effective.

A second manifestation of this method is in dataflow computing (see Dennis 1980; Veen 1986; Carlson 1985). Programs written for dataflow computers do not specify a sequence of execution. Rather, data elements flow from one processor to another as they are processed. Two instructions can be executed by different processors concurrently unless one requires the output of the other. Likewise, the execution of any given instruction is postponed until all the values it requires as input are available.

Either of these two moderately parallel approaches could improve the performance of existing generalization algorithms. In particular, a rule-based approach to generalization is likely to have many rules whose execution is, to some degree, independent of others. However, this approach does not offer a particularly novel outlook on algorithm design and would not purposefully desegregate the components of generalization. If the goal is to gain a new perspective on the task of generalization, other software models offer more.

Message passing

The message-passing model seals data structures within processors and establishes a formal mechanism (i.e. message passing) for communicating among different parallel activities. Message passing is associated with a coarse-grained, MIMD machine. Each parallel activity resembles a separate program using separate data structures (both of which, by implication, are executed by a separate processor). Programs operate on data structures to modify them. Because few problems break into entirely independent parts, the model permits one part to pass messages to another at strategic moments. The fact that a processor volunteers messages to other processors, but cannot peek into another processor's memory, reduces the risk that one processor will use incomplete computations from another as input. This model overcomes contention for shared resources by sharing no memory, and by rigidly controlling the channels of communication between processors.

A major difficulty of message passing is knowing what messages to generate and where to send them. For this reason, the message-passing approach is most appropriate for problems that break into large and relatively independent units. Gelernter (1987) cites the example of simulating planetary movements and gravitational pulls. For cartographic generalization, a message-passing approach could be implemented by performing first functional, then domain decomposition. If the algorithms to perform each component of generalization were parcelled out among processors, a band-sweep approach to processing the space could pass data among processors until an area became stable (i.e. no more generalization was needed), at which point it would be flushed from the processors and written to storage (Fig. 12.1). Alternatively, processing could begin in one corner of the region and radiate outward, as have the tesselation algorithms of Brassel (1975) and Green and Sibson (1979).

A message-passing approach to generalization implemented with both functional and domain decomposition seems a reasonable approach to the desegregation problem. Although processing is not precisely concurrent, it mimics the the human's iterative examination and modification of an area to be generalized. Many questions arise; for example, precisely when should the algorithm flush data from the processors, how much duplicate data would the three processors need to retain, and how separable are the algorithms that are executed by each processor? In short, this approach presents tantalizing possibilities, but must be developed further before its potential is known.

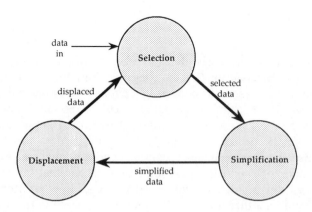

Fig. 12.1 The message-passing approach implemented with one component of generalization per processor

Connectionism

The alternative to message passing is to share one set of data structures among many processors, thereby reducing the need for processors to communicate among themselves. Two major manifestations of this model exist. The first is associated with a specific computer: the Connection Machine (Hillis 1985). The Connection Machine is a fine-grained computer that uses a SIMD/data-parallel approach. Memory is divided into small units, each containing a simple processor and a single data element. An instruction stream directs each processor to process the data contained in its memory; the actual transformation of data then occurs in parallel.

Connectionist algorithms differ markedly from the Von Neumann approach. Consider how to determine which children in a classroom had a hamburger for lunch. A Von Neumann approach would ask each child about his lunch in turn; a connectionist approach would request a show of hands and comprehend the location and number of raised hands instantaneously. Evidently, connectionism excels at some tasks but also requires a new outlook on old problems.

The usefulness of connectionism to raster generalization is apparent (Hillis 1985 uses image processing as an exemplary application of connectionism). It is less obvious how to use this powerful software model in the heterogeneous vector environment. One implementation would assign small geographical units to processors, then shift the grid slightly or recombine the units hierarchically as processing proceeds (Fig. 12.2(a) and (b)). A hierarchical approach is common for parallel routines (e.g. Hillis and Steele's 1986 algorithm for computing the sum of an array and the Bitton *et al.* 1983 example of a parallel binary merge sort).

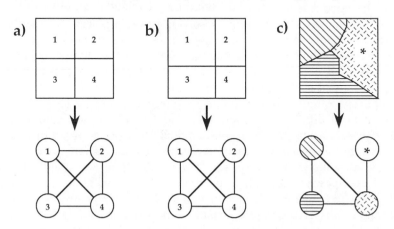

Fig. 12.2 A connectionist approach to transforming spatial data: (a) units of space are allotted to processors; (b) boundaries between units of space are shifted to overcome boundary effects; (c) features or thematic units are stored, one to a processor. Linkages between processors represent true topological relations

Alternatively, one could take a feature-based approach, where each processor would hold a feature and connections between processors would mimic real-world topology (Fig. 12.2(c)). The instruction stream would then broadcast rules, computations, or algorithms to transform these data. Since individual processors can pass values or combine themselves, many options exist for generalization in a connectionist environment. Connectionism provides a remarkable opportunity to gain a new perspective on spatial problems. But the adaptations of connectionism described here do not achieve the stated goal of this study: to integrate the components of generalization. Data are transformed in parallel, but functions are performed singly.

Parallel 'tuple space'

A second means of sharing data structures among processors is to permit a processor to lay exclusive claim to the data it is transforming for the duration of that transformation. This approach is based on the construct of 'tuple space', which is a hardware-independent abstraction that contains interactive objects called tuples (i.e. ordered sets). A new language, Linda, has been developed (Gelernter 1987) to demonstrate these concepts.

Tuple space includes active and passive tuples. Active tuples contain processes, which they execute on passive tuples containing data. Gelernter likens passive tuples to notes tacked on a bulletin board; a process untacks the note it needs, changes it as necessary, then replaces it on the board if it is still usable. An active tuple can consume or spawn passive tuples, and becomes a passive tuple when its process is complete. Armstrong (1991, Ch. 5 this volume) describes a method introduced by Holland *et al.* (1986) where a flock of rules competes for relevance in solving a given problem. As described by Armstrong, rules are treated more like suggestions than commands, until sufficient evidence builds to take action. Competing hypotheses could be identified as such and fired simultaneously in a tuple-space environment. The implications of such a method are exciting yet uncertain.

The tuple-space model is well suited to heterogeneous problems. Tuples can store data of different sizes and types, and program logic of different sorts. For these reasons, the tuple-space model seems appropriate to vector data generalization because of the heterogeneous data structures involved. To harness this model for purposes of generalization, we could allot a feature to each passive tuple, then unleash a set of active tuples containing generalization functions on the feature data. Alternatively, passive tuples could contain units of geographical space, subdivided according to acreage or data density. In either case, each active tuple would contain a generalization function to modify data in the passive tuples. An advantage of this approach is that active tuples are free to go where they are needed, meaning that processors can focus on areas that are particularly overcrowded once less congested regions are generalized.

Implications for future research

Several interesting points arise from this exploratory discussion, and it is possible to draw some preliminary conclusions about parallelism and cartography.

Task decomposition

Domain decomposition is an important technique to use for spatial data processing because of the amount of data involved and because it may provide relief from some age-old problems of spatial data structuring. However, if the researcher seeks to break away from one-task-at-a-time processing and emulate the holistic human approach, some functional decomposition is also important.

Concurrency

If functions are to be truly concurrent, duplicate data must be stored in the processors that perform the functions and the resulting changes to the data passed from one processor to another as updates. The luxury of multiplying data volume by a factor of the number of functions used is somewhat extravagant. It is better, perhaps, to think in terms of integration and iteration when attempting to replicate human actions, rather than to attempt to render the actions fully concurrent.

Parallel spatial software models

None of the software models that were explored here were developed with spatial information needs specifically in mind, although all are adaptable to that end. These points lead to some relatively untrod research areas, which are summarized below in a series of questions.

1. Is it possible to map characteristics of problems to characteristics of software models? Is it possible to generalize good and bad parallel approaches for different classes of geographical problems?
2. Can parallelism help us emulate the human cartographer's approach to map compilation (assuming that this is our goal)? While parallelism seems assured of offering improved data processing performance, what precisely are the theoretical benefits of parallel approaches?
3. What would a parallel software model designed specifically for geographical data processing look like? Would one model suffice? What would be its primary characteristics?

Inevitably, academic cartographers inhabit two worlds: a philosophical one and a technical one. The technical world is results-oriented, and accepts good results with no further questions. This pragmatic view is healthy for the discipline; it has produced a viable first-generation capability for auto-mated mapping and geographical information processing. In turn, this capability primes the public for newer and better developments, and inspires many with good minds to educate themselves and contribute to the discipline. But cartography also has a philosophical and theoretical side, which protests that getting good results most of the time is not enough. It is also important to define the nature of 'good results' and the methods used to achieve them, to explore alternative formats and processes, and to challenge our own assumptions. Adherence to the Von Neumann software model is a fundamental assumption that permeates cartography, and which will not be easy to overcome. Even if the rewards of challenging this assumption are limited to a new perspective on old problems, the rewards are worthwhile.

Acknowledgements

Much of the research and many of the thoughts expressed here originated during a period of digression as I was trying to complete my dissertation. I would like to thank Nick Chrisman for joining me in the digression, as well as for encouraging me to abandon it until a more opportune time; and also Barbara Buttenfield and Bob McMaster for creating that opportune time and giving me a forum in which to air these thoughts. Finally, I would like to thank Mary Clawson for a dinner conversation that led me in some new directions.

13

Integration and evaluation of map generalization

William A. Mackaness

Introduction

It is difficult to design maps by applying various generalization techniques in isolation of one another. The interdependent nature of simplification, selection, classification, symbolization, and the other techniques familiar in the cartographic process are, in the computer environment, often rendered one at a time, and the map design proceeds in terms of data type (lines, points, areas, text), without assessing the relative importance of information. To consider the question 'Is this item salient or merely contextual?' requires that map purpose is known, in what environment the map will be used, and by whom. Additionally it is necessary to know the spatial juxtaposition of items on a map. For example, a feature falling in a relatively sparse cultural area has a greater need to be preserved than an item in a map area dense with features. An appropriate mix of generalization techniques can be determined only intuitively, perhaps by viewing the map in various degrees of generalized detail. To this end one might propose the design of an interactive device to gauge the degree of generalization as one views the resulting map product, providing a real-time 'yardstick of goodness' by which to evaluate the application of each of the techniques. Before presenting the technique it will help the reader to provide some contextual background.

A subset of generalization techniques

The classification of generalization techniques is an academic process and the cartographer is not consciously aware of such distinctions when designing a map. Models for generalization have been published by several authors (e.g. Brassel and Weibel 1988; McMaster and Shea 1988; Buttenfield 1984; Ratajski 1967; Pannakoek 1962) and are reviewed and compared (McMaster 1991) in Chapter 2 in this volume. Many of the published models attempt a comprehensive taxonomy of all generalization operators, in sequential lists or hierarchic categorization. The alternative approach focuses upon process and structure to delineate a cartographic niche for specific tasks. There will be no attempt to add a new model to these efforts but instead the current discussion will draw upon these taxonomies to explain the principle of interactive evaluation of map generalization.

For the purpose of argument, let the taxonomy be limited to a subset of all possible generalization techniques, as illustrated in Fig. 13.1. Three techniques modify groups of symbols (change symbols, mask symbols, increase size differentiation between symbols). Three other techniques reduce the overall amount of detail (select, omit, simplify) and the last three reorganize but do not necessarily eliminate information (combine (reclassify), displace and exaggerate). For reasons of brevity, the author will consider the nine generalization techniques as a representative subset, and refer to them as a group throughout the chapter.

The complexities and nuances of map design are complex. Though there is much literature on the subject of specific generalization tasks (Nickerson and Freeman 1986; Freeman and Ahn 1984; Töpfer and Pillewizer 1966), to date only crude techniques have been developed for integrating operations. The exciting (and frustrating) thing about generalization is that the problem can be easily explained or illustrated using finished examples; and the human eye coupled with the artistic cartographic hand provides an immediate solution. Yet formal rules governing these intuitive decision-making processes remain elusive to automated processing. That integration is a key component of design cannot be in dispute (Karssen 1980; Tufte 1983). The reason for its omission under existing algorithms is perhaps due to its subjective and intangible nature (e.g. Tilghman 1984).

The key to the problem seems to be that it is the relative importance of features and their position on the map that governs the way in which they can be generalized. This phenomenon is referred to as **topogeneralization**. Topogeneralization is a phrase the author coined after reading an article on topobiology (Edelman 1989). Topobiology describes interactions among cells that depend on the cells' position. We can extend the analogy and define topogeneralization of features as the generalization of features dependent on the attributes as well as the relative position of features on the map. The inclusion or exclusion of a map feature is dependent on its relative importance and location with respect to other local features of similar character. Thus a grouping of buildings donates the idea of a town, and at the 'town level' features must compete for inclusion, similarly, but from the opposite viewpoint, scattered or isolated features (which express the impression of open tracts of land) are more critical in culturally sparse areas and therefore very likely to be included.

Such an approach to map design cannot be accommodated using current automated generalization techniques. The impact of graphical context on

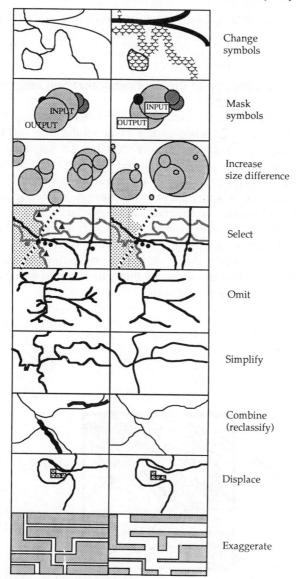

Change
symbols

Mask
symbols

Increase
size difference

Select

Omit

Simplify

Combine
(reclassify)

Displace

Exaggerate

Fig. 13.1 Nine generalization techniques

automated name placement has been discussed recently by Wu and
Buttenfield (1990), but no proposals for integrating context into existing
algorithms was presented. To develop a fully automated rule-base approach
to map generalization, there is a need to develop character analysis
algorithms (at what scale and under what conditions does a group of
buildings become a town?). One mechanism to evaluate the integration of
applied generalization operators can be readily implemented given existing
graphical user interfaces. Implementation of this concept could introduce

topogeneralization as a viable evaluation technique for interactive map generalization.

Integrating the application of the nine generalization techniques

The approach taken here accommodates links between map functionality and description with appropriate choice of generalization techniques. The solution integrates many of the factors that human cartographers would expect to have at their fingertips. The same factors govern the mixture and tolerance of the various generalization techniques when applied in a manual map design procedure. Figure 13.2 shows a portion of a topographic map at scale 1 : 50 000, which can be imagined as a compilation of a series of layers of data, for example, a transport layer, land use layer, elevation layer, etc. In its current state it is of little use (irrespective of map task) as it is too congested with information. Many unique combinations of generalization techniques can be applied in varying amounts to different groups of features to convert the map into a usable state.

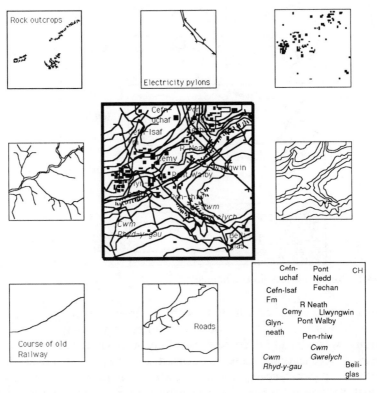

Fig. 13.2 Layers of topographic map information at scale 1 : 50 000. One might think of this as a plot of all the information in all layers of a topographic database

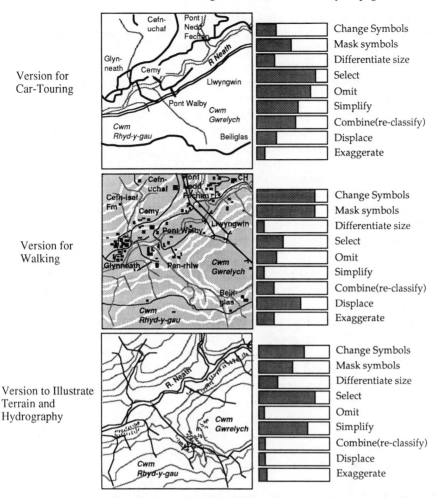

Fig. 13.3 Three combinations of generalization operations for specific map purposes

To illustrate, Fig. 13.3 shows three solutions, designed by hand. The histogram alongside each version is a qualitative indication of the degree of application of the generalization technique to the map as a whole. Each histogram bar is itself a summary of a single operator described in the previous section. For example, a short histogram indicates light application of the technique, as for example in the first solution where minimal symbol masking, size differentiation and exaggeration have been applied. A longer bar (reselection, in the same solution) indicates a more intensive application. Notice that numeric indicators are not provided to the user, and therefore the power of the graphical tool is ordinal rather than metric. A metaphor used in several computer operating systems to describe these types of diagrams is 'thermometer' (Apple Computer 1987). The set of all histogram bars becomes a 'generalization thermometer' describing not only the application of an individual technique but also its relative intensity with respect to other operators.

The first solution was designed with road users in mind, the second (beneath the first) attempts to encapsulate more information pertinent to cross-country walkers and the third (at the bottom) emphasizes terrain and hydrology. Notice that each solution may require some refinement, as for example in the masking of place-names to clarify the second solution: it is important to recognize that the thermometer is not intended to produce optimal design, but rather to assist the cartographer in evaluating relative importance of individual operators as the optimal design is approached. The optimal design has been cited as the one that provides 'the greatest number of ideas in the shortest time, with the least ink in the smallest space' (Tufte 1983: 51); it may be generated using some unique blend of thermometers, and clearly must be evaluated in the context of a specific map purpose and audience.

Another order of complexity is thereby introduced into the problem of topogeneralization and interactive evaluation. The question is how can rules to produce good maps be derived that govern integration of the various generalization techniques? Current automated cartographic systems cannot accommodate the large number of variables that must be considered, nor the level of detail at which the design decision-making process operates. Nor can they distinguish and order the map design objectives. Another graphical technique might make the problem more manageable by incorporating principles of pattern recognition.

Rose diagrams

A rose diagram is a circular graph employing a series of polar axes to represent ordinal or metrically scaled phenomena, for example wind direction and speed. A phenomenon is measured on each of the axes, and lines connecting the measurements form an irregular star-shape (as in a wind-rose). The use of polar coordinate graphs is not new to geographical description. Geographers use rose diagrams for a number of applications, for example rock orientations may be illustrated to determine direction of glacial movement (Verhoogen *et al.* 1970). Food scientists use them for determining the signatures of foods (Bennion 1980). The earliest diagrams were used for classifying beer.

Figure 13.4 shows one for classifying wine. It enables the encapsulation of the senses (touch, smell, taste, sight) into one diagram. Ratings may be provided metrically by using semantic scaling methods (Osgood, Suci, and Tannenbaum 1957); with this method, one could add standard deviation markers to the average ratings given on each axis. The evaluation of various types of wine is converted to one of shape recognition. Other wines

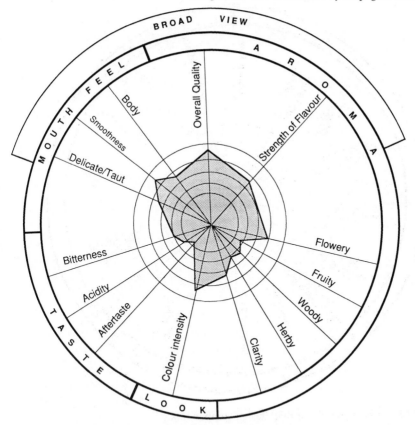

Fig. 13.4 A typical rose diagram describing wine in sensory terms

whose taste is similar to the illustration should have rose diagrams of similar shape. Differences in shape can be quickly identified as differing ratings for a particular aspect of the wine taste, for example acidity or sweetness. That is, the visual tool (shape recognition) provides quick evaluation of relative contributions of each rating, as well as providing quick comparisons between wines.

Rose diagrams may also prove valuable in evaluating a series of map generalization solutions. The spikes measure map variables that represent the anticipated map requirements. Petchenik (1974) selects a hierarchy of map design aspects, including physical measurements (map size, percentage inked, etc.), sensations (of colour or other non-emotional impressions), specific and general impressions (of like or dislike, for example) and finally associations (emotional responses to the map image). The goal is to incorporate rankings of map type or function, map purpose, and the map user (the intended audience). Application of semantic axes to the map generalization problem discussed early in this chapter would not emphasize emotional responses so much as the applicability of the map generalization solution to a particular map task.

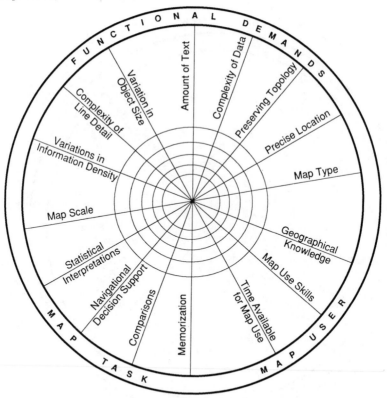

Fig. 13.5   Rose diagram for map design components

Figure 13.5 represents one possible configuration for a map generalization rose diagram. The reader should be cautioned that the rose diagrams do not represent the themometer bars directly. What is illustrated are map reader expectations and responses. Such a rose diagram could be generated for a road map, or a hiking map, or a map emphasizing terrain and hydrography. In each case, one would expect the shape of the diagram to vary with map reader expectations about the level of complexity, optimal map scale, and the amount of time available for map reading. Though not considered comprehensive, all the included attributes contribute to the map generalization process. In the case of time available for reading the map for example, shorter fixation periods require better contrast to readily distinguish features; greater emphasis would therefore need to be given to place-name labelling.

The utility of rose diagrams for integrating various generalization techniques lies in the formation of relationships between the star-shape describing map reader expectations about a particular mapping situation and the thermometers integrating operators to produce one or more generalized solutions. Figure 13.6 illustrates this for the first two solutions given in Fig. 13.3, together with their thermometers. Relationships between the rose diagrams and the thermometers must be derived by empirical examination,

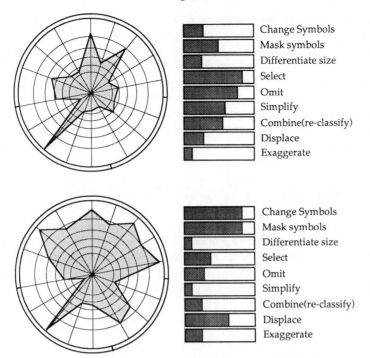

	Change Symbols
	Mask symbols
	Differentiate size
	Select
	Omit
	Simplify
	Combine(re-classify)
	Displace
	Exaggerate

	Change Symbols
	Mask symbols
	Differentiate size
	Select
	Omit
	Simplify
	Combine(re-classify)
	Displace
	Exaggerate

Fig. 13.6   Rose diagrams and thermometers comparing first two generalization solutions for the car touring map (top) and the walking map (bottom). These two maps are illustrated in Fig. 13.3

of course; what is presented here is a proposal rather than an implementation. Research questions to uncover these relationships might centre on a variety of topics. For example, on what criterion should the selection of rose diagram axes be based? What specific links can be determined between scaling of specific axes and the intensity with which specific generalization techniques are applied? For example, does simplification always reduce map complexity? And finally, if the rose diagrams can be linked with the thermometers, is it possible to make that link interactive, so that the shape of a rose diagram can be seen to vary in real time as one drags the bars of the thermometer to adjust the map image?

Summary

Current generalization techniques are not applied in isolation in a manual context, and their integration remains an impediment to fully automated mapping. The graphical context of map items requires understanding in order to optimally integrate the various generalization operations. There is, in addition, a need to attend to issues of topogeneralization, or relevance of proximal features, in generalizing a map. Graphical tools to illustrate methods for blending and integrating of generalization techniques have been suggested. The thermometer provides a visual tool by which to monitor the application of techniques, and the rose diagram provides guidelines for map reader expectations for a given map purpose. The rose diagrams facilitate map comparisons in that they exploit the visual activity of humans for shape recognition and shape comparison. Determination of relations between the rose diagram axes (map attributes) and the thermometers (showing the mix of generalization techniques) remains an unresolved issue. Empirical investigation may provide the most efficient mechanism to uncover these relationships.

Optimal map design (i.e. the successful selection of generalization techniques which includes the selection of salient and contextual data) will depend on a clear understanding of complex spatial relationships (such as is discussed by Peuquet 1988b; Palmer and Frank 1988), as well as an understanding of how map images are interpreted (Schlictmann 1984). Data representations must reflect the principles of effective visualization (Robertson 1988) and 'until visual character can be analytically identified, it will remain a major stumbling block in automating generalization tasks' (Buttenfield 1985: 20).

Acknowledgements

The author wishes to acknowledge the financial assistance of the Royal Institute of Chartered Surveyors' Education Trust Award from the AUTO-CARTO LONDON Funds. This chapter reports research that is part of NCGIA Initiative 3, Multiple Representations and is supported in part by a grant from the National Science Foundation (SES 88-10917); support from NCGIA during a postdoctoral residence at SUNY-Buffalo is appreciated.

References

Anderson J R, Hardy E E, Roach J T, Witmer R E 1976 *A Land Use and Land Cover Classification System for Use with Remote Sensor Data* Geological Survey Professional Paper 964, United States Government Printing Office, Washington, DC

Apple Computer 1987 *Human Interface Guidelines: The Apple Desktop Interface* Apple Computer, Inc, Cupertino, California

Armstrong M P 1991 Knowledge classification and organization. In Buttenfield B P, McMaster R B (eds) *Map Generalization: Making Rules for Knowledge Representation* Longman, London pp 86–102

Armstrong M P, Bennett D A 1990a A bit-mapped classifier for groundwater quality assessment. *Computers and Geosciences* 1(1): 811–32

Armstrong M P, Bennett D A 1990b A knowledge based object-oriented approach to cartographic generalization. *Proceedings GIS/LIS '90* Anaheim, California pp 48–57

Armstrong M P, De S, Densham P, Lolonis P, Rushton G, Tewari V 1990 A knowledge-based approach for supporting locational decision-making. *Environment and Planning B: Planning and Design* 17: 341–64

Armstrong M P, Lolonis P 1989 Interactive analytical displays for spatial decision support systems. *Proceedings AUTO-CARTO 9, Ninth International Symposium on Computer-Assisted Cartography* Baltimore, Maryland March 1989 pp 171–80

Attneave F 1954 Some informational aspects of visual perception. *Psychological Review* 61: 183–93

Backus J 1978 Can programming be liberated from the Von Neumann style? A functional style and its algebra of programs. *Communications of the Association for Computing Machinery* 21(8): 613–41

Ballard D 1981 Strip-trees: a hierarchical representation for curves. *Communications of the Association for Computing Machinery* 14: 310–21

Ballard D, Brown C 1982 *Computer Vision* Prentice-Hall, New York

Barr A, Feigenbaum E A 1981 *The Handbook of Artificial Intelligence* vol 1 William Kaufmann, Inc, Los Altos, California

Bartsch H J 1974 *Handbook of Mathematical Formulas* Academic Press, New York

Beard M K 1987 How to survive on a single detailed database. *Proceedings AUTO-CARTO 8, Eighth International Symposium on Computer-Assisted Cartography*, Baltimore, Maryland March 1987 pp 211–20

Beard M K 1988 Multiple Representations from a detailed database: a scheme for automated generalization. Unpublished PhD thesis, University of Wisconsin, Madison, Wisconsin 322 pp.

Beard M K 1991 Constraints on rule formation. In Buttenfield B P, McMaster R B (eds) *Map Generalization: Making Rules for Knowledge Representation* Longman, London pp 121–35

Bennion M 1980 *The Science of Food* John Wiley and Sons, New York

Berry J K 1987 Fundamental operations in computer-assisted map analysis. *International Journal of Geographical Information Systems* **1**: 119–36

Bertin J 1973 *Sémiologie graphique* 2nd edn Gauthier-Villars, Paris

Bertin J 1983 *Semiology of Graphics* University of Wisconsin Press, Madison, Wisconsin

Bitton D, Boral H, DeWitt D J, Wilkinson K W 1983 Parallel algorithms for the execution of relational database operations. *Association for Computing Machinery Transactions on Database Systems* **8**(3): 324–53

Board C 1984 Higher order map-using tasks: geographical lessons in danger of being forgotten. *Cartographia* **21**(1): 85–97

Booker L B, Goldberg D E, Holland J H 1989 Classifier systems and genetic algorithms. *Artificial Intelligence* **40**: 235–82

Boyle A R 1970 The quantised line. *The Cartographic Journal* **7**(2): 91–4

Brachman R J 1979 On the epistemological status of semantic networks. In Findler N V (ed) *Associative Networks: Representation and Use of Knowledge by Computers* Academic Press, New York pp 3–50

Brassel K E 1975 Neighborhood computations for large sets of data points. *Proceedings AUTO-CARTO 2, Second International Symposium on Computer-Assisted Cartography* Reston, Virginia Sept 1975 pp 337–45

Brassel K E 1985 Strategies and data models for computer-aided generalization. *International Yearbook of Cartography* **25**: 11–29

Brassel K E 1990 *Kartographisches generalisieren* Schweizerischen Gesellschaft für Kartographie Publikationen, Zurich, Switzerland

Brassel K E, Weibel R 1987 Map generalization. In Anderson K E, Douglas A V (eds) *Report on International Research and Development in Advanced Cartographic Technology, 1984–1987* International Cartographic Association, pp 120–37

Brassel K E, Weibel R 1988 A review and conceptual framework of automated map generalization. *International Journal of Geographical Information Systems* **2**(3): 229–44

Brodie M 1984 On the development of data models. In Brodie M, Mylopoulos J, Schmidt J (eds) *On Conceptual Modeling: Perspectives from Artificial Intelligence, Database and Programming Languages* Springer-Verlag, New York pp 19–48

Brophy D M 1972 Automated linear generalization in thematic cartography. Master's thesis, Department of Geography, University of Wisconsin, Madison, Wisconsin

Brown J H 1987a Development of prototype procedure 1 : 250 000 project. Internal report, Surveys and Mapping Branch, Energy, Mines and Resources, Ottawa, Canada Jan 1987

Brown J H 1987b Final report 1 : 250 000 project. Internal report, Surveys and Mapping Branch, Energy, Mines and Resources, Ottawa, Canada March 1987

Brownston L, Farrell R, Kant E, Martin N 1985 *Programming Expert Systems in OPS5: An Introduction to Rule-Based Programming* Addison-Wesley, Reading, Massachusetts

Burrough A 1986 *Principles of Geographical Information Systems for Land Resource Assessment* Oxford University Press, Oxford

Buttenfield B P 1984 Line structure in graphic and geographic space. Unpublished PhD dissertation, Department of Geography, University of Washington, Seattle, Washington

Buttenfield B P 1985 Treatment of the cartographic line. *Cartographica* **22**(2): 1–26

Buttenfield B P 1991 A rule for describing line feature geometry. In Buttenfield B P, McMaster R B (eds) *Map Generalization: Making Rules for Knowledge Representation* Longman, London pp 150–71

Buttenfield B P 1987 Automating the identification of cartographic lines. *The American Cartographer* **14**(1): 7–20

Buttenfield B P 1989 Scale dependence and self-similarity of cartographic lines. *Cartographica* **26**(1): 79–100

Buttenfield B P 1991 A rule for describing line feature geometry. In Buttenfield B P, McMaster R B (eds) *Map Generalization: Making Rules for Knowledge Representation* Longman, London pp 150–71

Buttenfield B P, DeLotto J S 1989 *Multiple Representations: Report on the Specialist Meeting* NCGIA Report 89–3, National Center for Geographic Information and Analysis, Santa Barbara, California

Buttenfield B P, Mark D M 1991 Expert systems in cartographic design. In Taylor D R F (ed) *Geographic Information Systems: The Computer and Contemporary Cartography* Pergamon Press, Oxford pp 129–50

Caldwell D R, Zoraster S, Hugus M 1984 Automating generalization and displacement lessons from manual methods. *Technical Papers, 44th Annual American Congress on Surveying and Mapping (ACSM) Meeting* Washington, DC March 1984 pp 254–63

Carlson W W 1985 Algorithmic performance of dataflow multiprocessors. *Computer* Dec: 30–40

Carpenter L C 1981 Computer rendering of fractal curves and surfaces. *Proceedings, Association for Computing Machinery (ACM) SIGGRAPH Conference* Seattle, Washington, **14**(3): 190 (abstract only). Text in *Computer Graphics* July 1980: 9–15

Catlow D, Du D 1984 The structuring and cartographic generalization of digital river data. *Technical Papers, 44th Annual American Congress on Surveying and Mapping (ACSM) Meeting* Washington, DC March 1984 pp 511–20

Child J 1984 A semiotic approach to cartographic structure and map meaning. Unpublished PhD dissertation, Department of Geography, University of Washington, Seattle, Washington

Chrisman N R 1983 Epsilon filtering: a technique for automated scale changing. *Technical Papers, 43rd Annual American Congress on Surveying and Mapping (ACSM) Meeting* Washington, DC March 1983 pp 322–31

Christ F 1978 A program for the fully automated displacement of point and line features in cartographic generalization. *Nachrichten aus dem Karten- und Vermessungswesen* Series 2, **35**: 5–30

Clarke K C 1990 *Computer and Analytical Cartography* Prentice-Hall, Englewood Cliffs, New Jersey

Daniels J 1986 Cognitive models in information retrieval – an evaluative review. *Journal of Documentation* **42**: 272–304

Dayal U, Smith J M 1986 PROBE: a knowledge-oriented database management system. In Brodie M L, Mylopoulos J (eds) *On Knowledge Base Management Systems* Springer-Verlag, New York pp 227–57

De S 1988 Knowledge representation in manufacturing systems. In Kusiak A (ed), *Expert Systems: Strategies and Solutions in Manufacturing Design and Planning* Michigan Society of Manufacturing Engineers, Dearborn pp 79–107

DeLucia A, Black T 1987 A comprehensive approach to automatic feature generalization. *Proceedings, 13th International Cartographic Association Conference* Morelia, Mexico 12–21 Oct 1987 **4** pp 168–91

Dennis J B 1980 Data flow supercomputers. *Computer* Nov: 48–56

Dent B 1990 *Cartography: Thematic Map Design* 2nd edn W C Brown Publishers, Dubuque, Iowa

Dettori G, Falcidieno B 1982 An algorithm for selecting main points on a line.

Computers and Geosciences **8**(1): 3–10

Deveau T J 1985 Reducing the number of points in a plane curve representation. *Proceedings AUTO-CARTO 7, Seventh International Symposium on Computer-Assisted Cartography* Baltimore, Maryland March 1985 pp 152–60

DMA (Defense Mapping Agency) 1986 *Product Specifications for Digital Feature Analysis Data (DFAD), Level 1 and Level 2* 2nd edn Missouri DMA Aerospace Center, St Louis, April 1986

Doerschler J S, Freeman H 1989 An expert system for dense-map name placement. *Proceedings AUTO-CARTO 9, Ninth International Symposium on Computer-Assisted Cartography* Baltimore, Maryland March 1989 pp 215–24

Douglas D H, Peucker T K 1973 Algorithms for the reduction of the number of points required to represent a line or its caricature. *The Canadian Cartographer* **10**(2): 112–23

Duecker K J, Kjerne D 1987 Application of the object-oriented paradigm to problems in geographic information systems. *Proceedings, International Geographic Information Systems (IGIS) Symposium: The Research Agenda* Arlington, Virgina Nov 1987 **2** pp 79–87

Edelman G M 1989 Topobiology. *Scientific American* **260**(5): 44–53

Egenhofer M, Frank A 1989 Object-oriented modeling for GIS: inheritance and propagation. *Proceedings AUTO-CARTO 9, Ninth International Symposium on Computer-Assisted Cartography* Baltimore, Maryland March 1989 pp 588–98

EMR (Energy, Mines, and Resources Canada) 1974 *Topographic Mapping Manual of Compilation Specifications and Instructions* 3rd edn Topographic Mapping Division, Surveys and Mapping Branch, Dept of Energy, Mines, and Resources, Ottawa

EMR 1989 *Derivation Guide* Canada Centre for Mapping, Dept of Energy, Mines and Resources, Ottawa, Canada

Engelbart D C 1962 *Augmenting Human Intellect: A Conceptual Framework* Stanford Research Institute, Summary Report for Contract AF 49(638)-1024 (also NTIS microfiche AD 289 565), Menlo Park, California

Feigenbaum E A 1977 The art of artificial intelligence: 1. themes and case studies of knowledge engineering. *Proceedings, 5th International Joint Conference on Artificial Intelligence, IJCAI-77* Cambridge, Massachusetts **2** pp 1014–29

Fikes R, Kehler T 1985 The role of frame-based representation in reasoning. *Communications of the Association for Computing Machinery* **28**: 904–20

Fisher P, Mackaness W A 1987 Are cartographic expert systems possible? *Proceedings AUTO-CARTO 8, Eighth International Symposium on Computer-Assisted Cartography* Baltimore, Maryland March 1987 pp 530–4

Flynn M J 1972 Some computer organizations and their effectiveness. *Industrial, Electronic and Electrical Engineering Transactions in Computing* **C33**(7): 592–603

Freeman H R, Ahn J 1984 AUTONAP – an expert system for automatic name placement. *Proceedings, 1st International Symposium on Spatial Data Handling* Zurich, Switzerland Aug 1984 pp 544–69

Frey W 1986 A bit-mapped classifier. *BYTE Magazine* **11**: 161–72

Gallaire H, Minker J, Nicolas J M 1984 Logic and databases: a deductive approach. *Computing Surveys* **16**: 153–85

Gardiner V 1982 Stream networks and digital cartography. *Cartographica* **19**: 38–44

Gelernter D 1987 Programming for advanced computing. *Scientific American* **257**(10): 91–8

Goldberg D E 1989 *Genetic Algorithms in Search, Optimization and Machine Learning* Addison-Wesley, Reading, Massachusetts

Goodchild M F 1980 Fractals and the accuracy of geographical measures. *International Journal of Mathematical Geology* **12**: 85–98

Goodchild M F 1987 Towards an enumeration and classification of GIS functions. *Proceedings, International Geographic Information Systems (IGIS) Symposium:*

The Research Agenda Arlington, Virgina Nov 1987 **2** pp 67–77

Gottschalk H J 1973 The derivation of a measure for the diminished content of information of cartographic line smoothed by means of a gliding arithmetic mean. *Information Relative to Cartography and Geodesy, Translations* **30**: 11–16

Graklanoff G J 1985 Expert system technology applied to cartographic process – considerations and possibilities. *Technical Papers, 45th Annual American Congress on Surveying and Mapping (ACSM) Meeting* Washington, DC March 1985 pp 613–24

Green J, Sibson Robert 1979 Computing Dirichlet tesselations in the plane. *The Computer Journal* **21**(2): 168–73

Gregory K J, Walling D E 1973 *Drainage Basin Form and Process* Edward Arnold, London

Grunreich D 1985 Computer-assisted generalization. *Papers; Cerco Cartography Course* Frankfurt, Germany

Guptill S C 1990 *An Enhanced Digital Line Graph Design* US Geological Survey Circular 1048, Reston, Virginia

Guptill S, Fegeas R 1988 Feature based spatial data models – the choice for global databases in the 1990's. In Mounsey H, Tomlinson R F (eds) *Building Databases for Global Science* Taylor & Francis, London pp 279–95

Hake G 1975 Zum Begriffssystem der Generalisierung. *Nachrichten aus dem Karten-und Vermessungswesen* Special Issue: 53–62

Harmon H, King D 1985 *Expert Systems* John Wiley, New York

Hart A 1986 *Knowledge Acquisition for Expert Systems* McGraw-Hill, New York

Hayes-Roth F, Waterman D A, Lenat D B 1983 *Building Expert Systems* Addison-Wesley, Reading, Massachusetts

Hillis W D 1985 *The Connection Machine* MIT Press, Cambridge, Massachusetts

Hillis W D, Steele G L 1986 Data parallel algorithms. *Communications of the Association for Computing Machinery* **29**(12): 1170–83

Hillsman E L 1984 The *P*-median structure as a unified liner model for location–allocation analysis. *Environment and Planning A* **16**: 305–18

Holland J H 1986 Escaping brittleness: the possibilities of general-purpose learning algorithms applied to parallel rule-based systems. In Michalski R S, Carbonell J G, Mitchell T M (eds) *Machine Learning: An Artificial Intelligence Approach* Morgan Kaufman, Los Altos pp 593–623

Holland J H, Holyoak K J, Nisbett R E, Thagard R 1986 *Induction: Processes of Inference, Learning and Discovery* MIT Press, Cambridge, Massachusetts

Hu D 1989 *C/C++ for Expert Systems* Management Information Sciences, inc, Portland, Oregon

Hudson D L 1990 Autonomous view states: materialized support for view update propagation. Unpublished PhD thesis, Department of Surveying Engineering, University of Maine, Orono, Maine

Hwang K, Chowkwanyan R, Ghosh J 1989 Parallel architectures for implementing artificial intelligence systems. In Hwang K, DeGroot D (eds) *Parallel Processing for Supercomputers and Artificial Intelligence* McGraw-Hill, New York

Imhof E 1962 Die Anordnung der Namen auf der Karte. *International Yearbook of Cartography* **2**: 93–128. English translation Imhof E 1975 Positioning names on maps. *The American Cartographer* **2**(2): 128–44

Imhof E 1965 *Kartographische Geländedarstellung* Walter de Gruyter, Berlin, Germany

Imhof E 1982 *Cartographic Relief Presentation* Walter de Gruyter, Berlin, Germany

Jackson M 1986 *Introduction to Expert Systems* Addison-Wesley, Reading, Massachusetts

Jasinski M J 1990 *The Comparison of Complexity Measures for Cartographic Lines* NCGIA Report 90-1, National Center for Geographic Information and Analysis, Santa Barbara, California

Jenks George F 1981 Lines, computers and human frailties. *Annals of the Association of American Geographers* **71**(1): 1–10

Jenks George F 1985 Linear simplification: how far can we go? Paper presented to the 10th Annual Meeting of the Canadian Cartographic Association, Fredericton, New Brunswick June 1985

Jenks G F 1989 Geographic logic in map design. *Cartographica* **26**: 27–42

Johannsen T 1973 A program for editing and for some generalizing operations for derivation of a small scale map from digitized data in 1 : 50 000 scale. *Nachrichten aus dem Karten- und Vermessungswesen* series 2 **30**: 17–22

Jones C B 1989 Cartographic name placement with Prolog. *Industrial Electronic and Electrical Engineering Society Computer Graphics & Applications* **9**: 36–47

Jones C B, Abraham I M 1986 Design considerations for a scale-independent cartographic data base. *Proceedings, 2nd International Symposium on Spatial Data Handling* Seattle, Washington Aug 1986 pp 384–98

Karssen A J 1980 Artistic elements in map design. *Cartographic Journal* **17**(2): 124–7

Keates J S 1973 *Cartographic Design and Production* John Wiley, New York

Keates J S 1989 *Cartographic Design and Production* Longman, London

Kelley S 1977 Information and generalization in cartographic communication. Unpublished PhD dissertation, Department of Geography, University of Washington, Seattle, Washington

Kim W, Lochovsky F (eds) 1989 *Object-oriented Concepts, Databases, and Applications* Addison-Wesley, Reading, Massachusetts

King R 1989 My cat is object-oriented. In Kim W, Lochovsky F (eds) *Object-oriented Concepts, Databases, and Applications* Addison-Wesley, Reading, Massachusetts pp 23–30

Kowalski R A 1979 *Logic for Problem Solving* North-Holland, New York

Kuipers B 1978 Modeling spatial knowledge. *Cognitive Science* **2**: 129–53

Lakoff G 1987 *Women, Fire, and Dangerous Things* University of Chicago Press, Chicago, Illinois

Lang T 1969 Rules for the robot draughtsmen. *The Geographical Magazine* **42**(1): 50–1

Langran G E 1991 Generalization and parallel computation. In Buttenfield B P, McMaster R D (eds) *Map Generalization: Making Rules for Knowledge Representation* Longman, London pp 205–17

Langran G E, Poiker T K 1986 Integration of name selection and name placement. *Proceedings, 2nd International Symposium on Spatial Data Handling* Seattle, Washington July 1986 pp 50–64

Leberl F L, Olson D, Lichtner W 1985 ASTRA – a system for automated scale transition. *Technical Papers, 45th Annual American Congress on Surveying and Mapping (ACSM) Meeting* Washington, DC March 1985 **1** pp 1–13

Leopold L B, Wolman M G, Miller J P 1964 *Fluvial Processes in Geomorphology* W H Freeman, San Francisco, California

Lichtner W 1978 Locational characteristics and the sequence of computer assisted processes of cartographic generalization. *Nachrichten aus dem Karten- und Vermessungwegen* series **35**: 65–75

Lichtner W 1979 Computer-assisted processes of cartographic generalization in topographic maps. *Geo-Processing* **1**: 183–99

Mackaness W A 1991 Integration and evaluation of map generalization. In Buttenfield B P, McMaster R B (eds) *Map Generalization: Making Rules for Knowledge Representation* Longman, London pp 218–27

Mackaness W A, Fisher P 1987 Automatic recognition and resolution of spatial conflicts in cartographic symbolization. *Proceedings AUTO-CARTO 8, Eighth International Symposium on Computer-Assisted Cartography* Baltimore, Maryland March 1987 pp 709–18

Mackaness W A, Fisher R F, Wilkinson G G 1986 Towards a cartographic expert system. *Proceedings AUTO-CARTO LONDON, International Symposium on Computer-Assisted Cartography* London, UK Sept 1986 pp 578–87

McMaster R B 1983 Mathematical measures for the evaluation of simplified lines on maps. Unpublished PhD dissertation, Department of Geography and Meteorology, University of Kansas, Lawrence, Kansas

McMaster R B 1986 A statistical analysis of mathematical measures for linear simplification. *The American Cartographer* **13**(2): 103–16

McMaster R B 1987 Automated line generalization *Cartographica* **24**(2): 74–111

McMaster R B 1989a Introduction to 'numerical generalization in cartography'. *Cartographica* **26**(1): 1–6

McMaster R B 1989b The integration of simplification and smoothing algorithms in line generalization. *Cartographica* **26**(1): 101–21

McMaster R B 1991 Conceptual frameworks for geographical knowledge. In Buttenfield B P, McMaster R B (eds) *Map Generalization: Making Rules for Knowledge Representation* Longman, London pp 21–39

McMaster R B, Monmonier M S 1989 A conceptual framework for quantitative and qualitative raster-mode generalization. *Proceedings GIS/LIS '89* Orlando, Florida **2** pp 390–403

McMaster R B, Shea K S 1988 Cartographic generalization in a digital environment: a framework for implementation in a geographic information system. *Proceedings GIS/LIS '88* San Antonio Texas **1** pp 240–9

Maling D H 1968 How long is a piece of string? *Cartographic Journal* **5**: 147–56

Mandelbrot B B 1967 How long is the coast of Britain? Statistical self-similarity and fractal dimension. *Science* **156**: 636–8

Mandelbrot B B 1986 Self-affine fractal sets. In Pietrano L, Tossato E (eds) *Fractals in Physics* Elsevier, Amsterdam

Manola F, Brodie M L 1986 On knowledge-based system architectures. In Brodie M L, Mylopoulos J (eds) *On Knowledge Base Management Systems* Springer-Verlag, New York pp 34–54

Marino J S 1979 Identification of characteristics along naturally occurring lines: an empirical study. *The Canadian Cartographer* **16**(1): 70–80

Mark D M 1979 Phenomenon-based data structuring and digital terrain modeling. *Geo-Processing* **1**: 27–36

Mark D M 1983 Relations between field surveyed channel networks and map-based geomorphometric measures in small basins near Inez, Kentucky. *Annals, Association of American Geographers* **73**: 358–72

Mark D M 1989 Conceptual basis for geographic line generalization. *Proceedings AUTO-CARTO 9, Ninth International Symposium on Computer-Assisted Cartography* Baltimore, Maryland March 1989 pp 68–77

Mark D M 1991 Object modelling and phenomenon-based generalization. In Buttenfield B P, McMaster R B (eds) *Map Generalization: Making Rules for Knowledge Representation* Longman, London pp 103–18

Mark D M, Aronson P B 1984 Scale-dependent fractal dimensions of topographic surfaces: an empirical investigation, with applications in geomorphology and computer mapping. *International Journal of Mathematical Geology* **16**(7): 671–83

Marr D 1982 *Vision* W H Freeman, San Francisco

Meyer B 1988 *Object-oriented Software Construction* Prentice-Hall, New York

Meyer U 1986 Software developments for computer-assisted generalization. *Proceedings AUTO-CARTO LONDON*, London UK Sept 1986 vol 2 pp 247–56

Meyer U 1987 Computer-assisted generalization of building for digital landscape models by classification methods. *Nachrichten aus dem Karten- und Vermessungswesen* Monograph 2(46)

Michaelsen R H, Michie D, Boulanger A 1985 The technology of expert systems. *BYTE Magazine* **10**: 303–12

Minsky M 1975 A framework for representing knowledge. In Winston P H (ed) *The Psychology of Computer Vision* McGraw-Hill, New York pp 211–77

References

Monmonier M 1976 Statistical maps, algorithms, and generable behavior. *Technical Papers, 36th Annual American Congress on Surveying and Mapping (ACSM) Meeting* Washington, DC March 1976 pp 407–12

Monmonier M 1982 *Computer-assisted Cartography: Principles and Prospects* Prentice-Hall, Englewood Cliffs, New Jersey

Monmonier M 1983 Raster-mode area generalization for land use and land cover maps. *Cartographica* **20**(4): 65–91

Monmonier M 1986 Toward a practicable model of cartographic generalization. *Proceedings AUTO-CARTO LONDON, International Symposium on Computer-Assisted Cartography*, London, UK Sept 1986 **2** pp 257–66

Monmonier M 1987 Displacement in vector- and raster-mode graphics. *Cartographica* **24**(4): 25–36

Monmonier M 1989a Caricature-weighted and caricature-screened line simplification: expert-guided approaches to cartographic line generalization. *Journal of the Pennsylvania Academy of Science* **63**: 105–12

Monmonier M 1989b Interpolated generalization: cartographic theory for expert-guided feature displacement. *Cartographica* **26**(1): 43–64

Monmonier M 1989c Regionalizing and matching features for interpolated displacement in the automated generalization of digital cartographic databases. *Cartographica* **26**(2): 21–39

Monmonier M 1990 Graphically encoded knowledge bases for expert-guided feature generalization in cartographic display sysems. *International Journal of Expert Systems* **3**: 65–71

Monmonier M 1991 Role of interpolation in feature displacement. In Buttenfield B P, McMaster R B (eds) *Map Generalization: Making Rules for Knowledge Representation* Longman, London pp 189–204

Monmonier M, McMaster R B 1990 The sequential effects of geometric operators in cartographic line generalization. *International Yearbook of Cartography* (forthcoming)

Monmonier M, McMaster R B 1991 Geometric operators and sequential effects in cartographic line generalization. *International Yearbook of Cartography* (forthcoming)

Morrison J L 1974 A theoretical framework for cartographic generalization with emphasis on the process of symbolization. *International Yearbook of Cartography* **14**: 115–27

Muehrcke P C 1986 *Map Use: Reading Analysis, and Interpretation* 2nd edn JP Publications, Madison, Wisconsin

Muller J C 1989 Theoretical considerations for automated map generalization. *ITC Journal* **3/4**: 200–4, Enschede, The Netherlands

Muller J C 1990 Rule-based generalization: potentials and impediments. *Proceedings, 4th International Symposium on Spatial Data Handling* Zurich, Switzerland Aug 1990 **1** pp 317–34

Muller J C, Johnson R D, Vanzella L R 1986 A knowledge-based approach for developing cartographic expertise. *Proceedings, Second International Symposium on Spatial Data Handling* Seattle, Washington July 1986 pp 557–71

Mylopoulos J 1980 An overview of knowledge representation. *Proceedings of the Workshop on Data Abstraction, Databases, and Conceptual Modeling* pp 5–12

NCDCDS (National Committee for Digital Cartographic Data Standards) 1988 The proposed standard for digital cartographic data. *The American Cartographer* **15**(1): 9–140

Nickerson B G 1988a Automated cartographic generalization for linear features. *Cartographica* **25**(3): 15–66

Nickerson B G 1988b Data structure requirements for automated cartographic generalization. *Proceedings of the Third International Seminar on Trends and Concerns of Spatial Sciences* Laval University, Quebec City, Canada June 1988

Nickerson B G 1991 Knowledge engineering for generalization. In Buttenfield B P,

McMaster R B (eds) *Map Generalization: Making Rules for Knowledge Representation* Longman, London pp 40–56

Nickerson B G, Freeman H 1986 Development of a rule-based system for automatic map generalization. *Proceedings, Second International Symposium on Spatial Data Handling* Seattle, Washington July 1986 pp 537–56

Nierstrasz O 1989 A survey of object-oriented concepts. In Kim W, Lochovsky F H (eds) *Object-oriented Concepts, Databases, and Applications* Addison-Wesley, Reading, Massachusetts pp 3–21

Nyerges T L 1980 Representing spatial properties in cartographic databases. *Technical Papers, 40th Annual American Congress on Surveying and Mapping (ACSM) Meeting* St Louis, Missouri Sept 1980 pp 29–41

Nyerges T L 1987 GIS research issues identified during a cartographic standards process: spatial data exchange. *Proceedings, International Geographic Information Systems (IGIS) Symposium: The Research Agenda* Arlington, Virginia Nov 1987 **1** pp 319–29

Nyerges T L 1991a Analytical map use. *Cartography and Geographic Information Systems* **18**(1): 11–22

Nyerges T L 1991b Geographic information abstractions: conceptual clarity for geographic modeling. *Environment and Planning A*, in press

Nyerges T L 1991c Representing geographical meaning. In Buttenfield B P, McMaster R B (eds) *Map Generalization: Making Rules for Knowledge Representation* Longman, London pp 59–85

O'Brien D 1988 Generalization of map area data through raster processing. Internal report, Surveys and Mapping Branch, Energy, Mines, and Resources, Ottawa, Canada

Opheim H 1982 Fast data reduction of a digitized curve. *Geo-Processing* **2**: 33–40

Osgood C E, Suci G, Tannenbaum P H 1957 *The Measurement of Meaning* University of Illinois Press, Urbana, Illinois

Palmer B L, Frank A U 1988 Spatial languages. *Proceedings, Third International Symposium on Spatial Data Handling* Sydney, Australia July 1988 pp 201–10

Pannekoek A J 1962 Generalization of coastlines and contours. *International Yearbook of Cartography* **2**: 55–74

Parsaye K, Chignell M, Khoshafian S, Wong H 1989 *Intelligent Databases: Object-oriented, Deductive Hypermedia Technologies* John Wiley, New York.

Peckham J, Maryanski E 1988 Semantic data models. *Association for Computing Machinery Computing Surveys* **20**(3): 153–89

Perkal J 1966 An attempt at objective generalization. Trans. W Jackowski. In Nystuen J (ed) *Discussion Paper 10*, Michigan Inter-University Community of Mathematical Geographers, Ann Arbor, Michigan

Petchenik B B 1974 A verbal approach to characterizing the look of maps. *The American Cartographer* **1**(1): 63–71

Peucker T K 1975 A theory of the cartographic line. *Proceedings AUTO-CARTO 2, Second International Symposium on Computer-Assisted Cartography* Reston, Virginia Sept 1975 pp 508–18

Peuquet D J 1988a Representations of geographic space: toward a conceptual synthesis. *Annals of the American Association of Geographers* **78**(3): 375–94

Peuquet D J 1988b Towards the definition and use of complex spatial relationships. *Proceedings, Third International Symposium on Spatial Data Handling* Sydney, Australia July 1988 pp 211–23

Pfefferkorn C, Burr D, Harrison D, Heckman B, Oresky C, Rothermal J 1985 ACES: a cartographic expert system. *Proceedings, AUTO-CARTO 7, Seventh International Symposium on Computer-Assisted Cartography* Baltimore, Maryland March 1985 pp 399–407

Pike R J 1988 Toward geometric signatures for geographic information systems *Proceedings, International Geographic Information Systems (IGIS) Symposium:*

References

The Research Agenda Arlington, Virginia Nov 1987 vol 3 pp 15–26

Powitz B M 1989 Computer-assisted generalization of traffic networks and buildings. *Proceedings, 14th International Cartographic Association Conference* Budapest, Hungary Aug 1989 p 112 (abstract only)

Quillian M R 1968 Semantic memory. In Minsky M (ed) *Semantic Information Processing* MIT Press, Cambridge, Massachusetts

Raisz E 1962 *Principles of Cartography* John Wiley, New York

Ramer U 1972 An interactive procedure for the polygonal approximation of plane curves. *Computer Graphics and Image Processing* **1**: 244–56

Ratajski L 1967 Phénomènes des points de généralization. *International Yearbook of Cartography* **7**: 143–51

Reumann K, Witkam A K P 1974 Optimizing curve segmentation in computer graphics. *Proceedings, International Computing Symposium* North-Holland Publishing Company, Amsterdam pp 467–72

Rhind D 1973 Generalization and realism within automated cartography. *The Canadian Cartographer* **10**(1): 51–62

Rhind D 1988 A GIS research agenda. *International Journal of Geographical Information Systems* **2**(1): 23–8

Richardson D E 1988 Database design considerations for rule-based map feature selection. *ITC Journal* **2**: 165–71

Richardson D E 1989 Rule based generalization for base map production. *Proceedings, Challenge for the 1990's: Geographic Information Systems Conference* Canadian Institute for Surveying and Mapping, Ottawa, Canada, pp 718–39

Richardson D E, Muller J C 1991 Rule selection for small-scale map generalization. In Buttenfield B P, McMaster R B (eds) *Map Generalization: Making Rules for Knowledge Representation* Longman, London pp 136–49

Richardson L F 1961 The problem of contiguity; an appendix to the statistics of deadly quarrels. *General Systems Yearbook* **6**: 139–87

Ripple W, Ulshoefer V 1987 Expert systems and spatial data models for efficient geographic data handling. *Photogrammetric Engineering and Remote Sensing* **53**: 1431–3

Ritter D F 1986 *Process Geomorphology* Wm C Brown, Dubuque, Iowa

Robertson P K 1988 Choosing data representations for the effective visualization of spatial data. *Proceedings, Third International Symposium on Spatial Data Handling* Sydney, Australia July 1988 pp 243–52

Robinson A H 1960 *Elements of Cartography* 2nd edn John Wiley, New York

Robinson A H, Sale R D 1969 *Elements of Cartography* 3rd edn John Wiley, New York

Robinson A H, Sale R D, Morrison J L 1978 *Elements of Cartography* 4th edn John Wiley, New York

Robinson A H, Sale R D, Morrison J L, Muehrcke P C 1984 *Elements of Cartography* 5th edn John Wiley, New York

Robinson G, Jackson M 1985 Expert systems in map design. *Proceedings AUTO/CARTO 7, Seventh International Symposium on Computer-Assisted Cartography* Baltimore, Maryland March 1985 pp 430–9

Robinson G J, Zaltash A 1989 Application of expert systems to topographic map generalisation. *Proceedings, Association for Geographic Information Conference (AGI 89)* A3 1–A3 6

Robinson V B, Thongs D, Frank A U, Blaze M 1986 Expert systems and geographic information systems: critical review and research needs. *Geographic Information Systems in Government* US Army Engineer Topographic Laboratories Reports **2**: 851–69

Saalfeld A 1988 Conflation: automated map compilation. *International Journal of Geographical Information Systems* **2**: 217–28

Sayer A 1984 *Method in social science* Hutchinson, London

236

Schlictmann H 1984 Discussion of C. Grant Head 'The map as a natural language: a paradigm for understanding'. *Cartographia* 21(1): **33–6**

Schorr H 1988 Expert systems: an IBM perspective. Proceedings, Eighth International Workshop on Expert Systems and their Applications Avignon, France May 1988 vol 1 pp 39–40 (abstract only)

Schrefl M, Tjoa A, Wagner R 1984 Comparison criteria for semantic data models. *Proceedings, Industrial Electronic and Electric Engineering Society Conference on Data Engineering* Los Angeles pp 120–5

Scott A J 1971 Combinatorial programming, spatial analysis and planning. Methuen, London, UK

Shea K S 1991 Design considerations for an artificially intelligent system. In Buttenfield B P, McMaster R B (eds) *Map Generalization: Making Rules for Knowledge Representation* Longman, London pp 3–20

Shea K S, McMaster R B 1989 Cartographic generalization in a digital environment: when and how to generalize. *Proceedings AUTO-CARTO 9, Ninth International Symposium on Computer-Assisted Cartography* Baltimore, Maryland March 1989 pp 56–67

Shokalsky U M 1930 *Dlina glavneyshikh rek aziatatskoy chasti SSSR I sposob izmereniya dlin rek po kartam* Rechtransizdat, Moscow (as discussed in Maling D H 1968 How long is a piece of string? *Cartographic Journal* 5(2): 147–56.)

Sinton D 1978 The inherent structure of information as a constraint to analysis: mapped thematic data as a case study. In Dutton G (ed) *Harvard Papers on GIS* Addison-Wesley, Reading, Massachusetts 7 pp 1–17

Smith J M, Smith D C P 1977 Database abstractions; aggregation and generalization. *Association for Computing Machinery Transactions, Database Systems* 2(2): 105–33

Smith T R 1984 Artificial intelligence and its applicability to geographical problem solving. *The Professional Geographer* 36(2): 147–58

Smith T R, Pellegrino J W, Golledge R G 1982 Computational process modeling of spatial cognition and behavior. *Geographical Analysis* 14: 305–25

Sowa J 1984 *Conceptual Structures: Information Processing in Mind and Machine* Addison-Wesley, Reading, Massachusetts

Spiess E 1990 Bemerkungen zu wissensbasierten Systemem für die Kartographie. *Vermessung, Photogrammetrie und Kulturtechnik* 2(90): 75–81

Sprague R, Carlson E 1982 *Building Effective Decision Support Systems* Prentice-Hall, Englewood Cliffs, New Jersey

Steinhaus H 1954 Length, shape and area. *Colloquium Mathematica* 3

Steinhaus H 1960 *Mathematical Snapshots* Oxford University Press

Sterling L, Shapiro E 1986 *The Art of Prolog* MIT Press, Cambridge, Massachusetts

Steward H J 1974 Cartographic generalization: some concepts and explanations. *Cartographica* Monograph 10, University of Toronto Press, Toronto

Stormark E, Bie S W 1980 Interactive map editing – a first or a last resort? In Opheim H (ed) *Contributions to Map Generalization* Norwegian Computing Centre Publication 679 Trondheim, Norway pp 59–66

Teitz M B, Bart M 1968 Heuristic methods for estimating the generalized vertex median of a weighted graph. *Operations Research* 16: 955–61

Thapa K 1988 Automatic line generalization using zero-crossing. *Photogrammetric Engineering and Remote Sensing* 54(4): 511–17

Thompson M M 1979 *Maps for America* USGS, Reston, Virginia

Tilghman B R 1984 *But is it art?* Blackwell, London

Tobler W R 1964 *An Experiment in the Computer Generalization of Maps* Technical Report 1, Office of Naval Research, ASAD Contract 459953, Dec 1964, Washington, DC

Tobler W R 1966 *Numerical Map Generalization* Michigan Inter-University Community of Mathematical Geographers Discussion Paper No 8, Ann Arbor, Michigan

Tomlin, C D 1983 Digital cartographic modeling techniques in environmental planning. Unpublished PhD thesis, School of Landscape Architecture, Yale University

Töpfer F, Pillewizer W 1966 The principles of selection, a means of cartographic generalization. *The Cartographic Journal* **3**(1): 10–16

Tufte E R 1983 *The Visual Display of Quantitative Information* Graphic Press, Cheshire, Connecticut

Ullman S 1985 Visual routines. In Pinker S (ed) *Visual Cognition* MIT Press, Cambridge, Massachusetts pp 97–159

USGS (US Geological Survey) 1964 *Instructions for Stereocompilation of Map Manuscripts Scribed at 1 : 24 000* Topographic Division, USGS March 1964, Reston, Virginia

USGS 1985 *Digital Line Graphs from 1 : 100 000 scale maps* USGS National Mapping Division, Reston, Virginia

Van Horn E K 1985 Generalizing cartographic databases. *Proceedings AUTO-CARTO 7, Seventh International Symposium on Computer-Assisted Cartography* Baltimore, Maryland March 1985 pp 532–40

Vanicek P, Woolnough D F 1975 Reduction of linear cartographic data based on generalization of pseudo-hyperbolae. *The Cartographic Journal* **12**(2): 112–19

Veen A H 1986 Dataflow machine architecture. *Association for Computing Machinery Computing Surveys* **18**(4): 365–96

Verhoogen J, Turner F J, Weiss L E, Wahrhaftig C, Fyfe W S 1970 *The Earth: An Introduction to Physical Geology* Holt, Rinehart & Winston, New York

Volkov N M 1949 Novyy sposob ismereniya dlin rek po kartam. *Izvestia AN SSSR, Seriya Geografiya i Geofizicheskaya* T **13**(2). As discussed in Maling D H 1989 *Measurements from Maps: Principles and Methods of Cartometry* Pergamon Press, Oxford

Wah B W, Li G 1989 Design issues of multiprocessors for artificial intelligence. In Hwang K, DeGroot D (eds) *Parallel Processing for Supercomputers and Artificial Intelligence* McGraw-Hill, New York

Waterman D A 1986 *A Guide to Expert Systems* Addison-Wesley, Reading, Massachusetts

Weibel R 1989a *Concepts and Experiments for the Automation of Relief Generalization* PhD Dissertation (in German), Geo-Processing Series No 15, Department of Geography, University of Zurich, Switzerland

Weibel R 1989b Design and implementation of a strategy for adaptive computer-assisted terrain generalization. *Proceedings, 14th International Cartographic Association Conference* Budapest, Hungary Aug 1989 pp 10–11 (abstract only)

Weibel R 1991 Amplified intelligence and rule-based systems. In Buttenfield B P, McMaster R B (eds) *Map Generalization: Making Rules for Knowledge Representation* Longman, London pp 172–86

Weibel R, Buttenfield B P 1988 Map design for geographic information Systems. *Proceedings, GIS/LIS '88* San Antonio, Texas vol 1 pp 350–9

Weiss S M, Kulikowski C A 1984 *A Practical Guide to Designing Expert Systems* Rowman and Allenheld, Totowa, New Jersey

White E R 1985 Assessment of line generalization algorithms using characteristic points. *The American Cartographer* **12**(1): 17–28

White M 1984 Technical requirements and standards for a multipurpose geographic data system. *The American Cartographer* **11**(1): 15–26

Wiederhold G 1986 Knowledge versus data. In Brodie M L, Mylopoulos J (eds) *On Knowledge Base Management* Springer-Verlag, New York pp 77–82

Wilson S 1981 Cartographic generalization of linear information in raster mode. Master's thesis, Department of Geography, Syracuse University, Syracuse, New York

Winograd T, Flores F 1987 *Understanding Computers and Cognition: A New Foundation for Design* Addison-Wesley, Reading, Massachusetts

Winston P H 1984 *Artificial Intelligence* 2nd edn Addison-Wesley, Reading, Massachusetts

Witkin A P 1986 Scale-space filtering. In Pentland A P (ed) *From Pixels to Predicates* Ablex Publishing Corp, Norwood, New Jersey pp 5–19

Wu C V, Buttenfield B P 1990 Reconsidering rules for point feature name placement. *Cartographia* **27**(4) (forthcoming)

Zoraster S 1986 Integer programming applied to the map label placement problem. *Cartographica* **23**(3): 16–27

Zoraster S, Davis D, Hugus M 1984 *Manual and Automated Line Generalization and Feature Displacement* US Army Engineering and Topographic Laboratory Report ETL-0359 Fort Belvoir, Virginia

Topical Index

Author Index